T0140010

# Advances in Intelligent Systems and Computing

Volume 562

**Series editor**

Janusz Kacprzyk, Polish Academy of Sciences, Warsaw, Poland
e-mail: kacprzyk@ibspan.waw.pl

## About this Series

The series "Advances in Intelligent Systems and Computing" contains publications on theory, applications, and design methods of Intelligent Systems and Intelligent Computing. Virtually all disciplines such as engineering, natural sciences, computer and information science, ICT, economics, business, e-commerce, environment, healthcare, life science are covered. The list of topics spans all the areas of modern intelligent systems and computing.

The publications within "Advances in Intelligent Systems and Computing" are primarily textbooks and proceedings of important conferences, symposia and congresses. They cover significant recent developments in the field, both of a foundational and applicable character. An important characteristic feature of the series is the short publication time and world-wide distribution. This permits a rapid and broad dissemination of research results.

## Advisory Board

Chairman

Nikhil R. Pal, Indian Statistical Institute, Kolkata, India
e-mail: nikhil@isical.ac.in

Members

Rafael Bello Perez, Universidad Central "Marta Abreu" de Las Villas, Santa Clara, Cuba
e-mail: rbellop@uclv.edu.cu

Emilio S. Corchado, University of Salamanca, Salamanca, Spain
e-mail: escorchado@usal.es

Hani Hagras, University of Essex, Colchester, UK
e-mail: hani@essex.ac.uk

László T. Kóczy, Széchenyi István University, Győr, Hungary
e-mail: koczy@sze.hu

Vladik Kreinovich, University of Texas at El Paso, El Paso, USA
e-mail: vladik@utep.edu

Chin-Teng Lin, National Chiao Tung University, Hsinchu, Taiwan
e-mail: ctlin@mail.nctu.edu.tw

Jie Lu, University of Technology, Sydney, Australia
e-mail: Jie.Lu@uts.edu.au

Patricia Melin, Tijuana Institute of Technology, Tijuana, Mexico
e-mail: epmelin@hafsamx.org

Nadia Nedjah, State University of Rio de Janeiro, Rio de Janeiro, Brazil
e-mail: nadia@eng.uerj.br

Ngoc Thanh Nguyen, Wroclaw University of Technology, Wroclaw, Poland
e-mail: Ngoc-Thanh.Nguyen@pwr.edu.pl

Jun Wang, The Chinese University of Hong Kong, Shatin, Hong Kong
e-mail: jwang@mae.cuhk.edu.hk

More information about this series at http://www.springer.com/series/11156

Ramesh K. Choudhary · Jyotsna Kumar Mandal
Dhananjay Bhattacharyya
Editors

# Advanced Computing and Communication Technologies

Proceedings of the 10th ICACCT, 2016

 Springer

*Editors*
Ramesh K. Choudhary
Asia Pacific Institute of Information
Technology
Panipat, Haryana
India

Dhananjay Bhattacharyya
Computational Science Division
Saha Institute of Nuclear Physics
Kolkata, West Bengal
India

Jyotsna Kumar Mandal
Department of Computer Science
and Engineering
Faculty of Engineering, Technology
and Management
Kalyani University
Kalyani, West Bengal
India

ISSN 2194-5357          ISSN 2194-5365   (electronic)
Advances in Intelligent Systems and Computing
ISBN 978-981-10-4602-5         ISBN 978-981-10-4603-2   (eBook)
https://doi.org/10.1007/978-981-10-4603-2

Library of Congress Control Number: 2017940607

Printed on acid-free paper

This Springer imprint is published by Springer Nature
The registered company is Springer Nature Singapore Pte Ltd.
The registered company address is: 152 Beach Road, #21-01/04 Gateway East, Singapore 189721, Singapore

# Preface

This AISC volume contains selected papers presented at the 10th International Conference on Advanced Computing and Communication Technologies (10th ICACCT 2016), technically sponsored by Institution of Electronics and Telecommunication Engineers (India), held during November 18–20, 2016 at Asia Pacific Institute of Information Technology, Panipat, India.

The technical program at 10th ICACCT 2016 featured inaugural addresses by Prof. Lalit M. Patnaik, National Institute of Advanced Studies, IISc, Bangalore and Mr. V. Ramagopal, Executive Director, Panipat Refinery and Petrochemicals, Indian Oil Corporation Ltd., Panipat; followed by six contributed paper sessions; Invited Paper Sessions and a valedictory session. The technical sessions were chaired by Prof. J.K. Mandal, Kalyani University; Prof. Dhananjay Bhattacharyya, Saha Institute of Nuclear Physics, Kolkata; Prof. Anu Mehra, AMITY University, Noida; Dr. Surya P. Singh, Indian Institute of Technology, Delhi; Mr. R. Ramamurthy, Dy. General Manager, Panipat Refinery and Petrochemicals, Indian Oil Corporation Ltd., Panipat; and Prof. Pradosh K. Roy, Asia Pacific Institute of Information Technology, Panipat.

Out of the contributed papers, 150 papers were shortlisted for second and third rounds of review by the expert panel after content evaluation and anti-plagiarism check using Turnitin®. After a rigorous review process, the final selection of 31 papers, corresponding to an acceptance rate of 20% was determined by these evaluation process. The selected papers cover a wide range of topics spanning theory, systems, applications, and case studies that could provide insights for using and enhancing advanced computing systems, communication technologies, services, and facilities.

A review of the state-of-the-art research work together with a discussion on CORDA as a computational model with reference to the problem of multi-robot area coverage has been enlightened in the invited paper.

The development of computational intelligence, it may be recalled, was inspired by observable and imitable aspects of intelligent activity of human being and nature. In the recent years, many hybrid techniques using the theories of computational intelligence are being applied in many fields of engineering and have many other practical implementations. Conforming to the theme of the conference, the volume includes systems based on computational intelligence on data of various nature, e.g., text mining, image thresholding and encryption, target coverage and localization in wireless sensor networks, dynamic facility layout, human activity recognition, software defect prediction, interactive recommender system, genes mediating leukemia, and glaucoma diagnosis for retinal images. The other chapters relate to knowledge representation and deep learning, innovations in data encryption, reliability in MANET, active filters, innovative antenna design, micro electromechanical (MEM) systems design, quantum dot cellular automata, wave digital filters, heterogeneous distributed computing systems, big data analysis and a seminal paper on statistical machine translation of Bodo language.

We would like to thank members of the advisory and the program committee, especially Prof. Nikhil Ranjan Pal, Indian Statistical Institute, Calcutta; Prof. Subir Sarkar, Jadavpur University, Calcutta; Prof. Partha P. Bhattacharya, Mody University, Sikar, Rajasthan; Prof. Alok K. Rastogi, Institute for Excellence in Higher Education, Bhopal; Prof. Siddhartha Bhattacharya, RCC Institute of Information Technology, Calcutta; Prof. Bijay Baran Pal, University of Kalyani, West Bengal; Prof. R.R.K. Sharma, Indian Institute of Technology, Kanpur; Prof. Sarmistha Neogy, Jadavpur University, Calcutta; Dr. Nitin Auluck, IIT Ropar, Rupnagar; Dr. Surya Prakash Singh, Indian Institute of Technology, Delhi; Dr. Srabani Mukhopadhyaya, BIT Mesra Kolkata Extension Centre, Calcutta; Dr. Mahavir Jhawar, Ashoka University, Sonepat, Haryana; Dr. Tandra Pal, National Institute of Technology, Durgapur; Prof. Atul Sharma, Thapar University, Patiala; Prof. Moirangthem Marjit Singh, Northeast Regional Institute of Science and Technology, Itanagar, Arunachal Pradesh, for their enormous contributions towards the success of this conference. The authors of all contributed papers, the session chairs, and the session coordinators deserve our special thanks. We are thankful to the authors of the selected papers for suitably incorporating the modifications suggested by the reviewers and the session chairs in the scheduled time frame.

Acceptance of the proposal submitted to Springer Science+Business Media Singapore Pte. Ltd. for publication of the proceedings is gratefully acknowledged. We are grateful to the Institution of Electronics and Telecommunication Engineers (India) for technically sponsoring the conference. We are indebted to Mr. Aninda Bose, Senior Editor, Springer India Pvt. Ltd., New Delhi, for his valuable advice on similarity index and the review process. Financial support from Panipat Refinery and Petrochemicals, IOCL, India, in organizing the conference is gratefully acknowledged.

The Organizing Committee patronized by the Managing Committee of APIIT SD India, viz., Mr. Vinod Gupta, Chairman; Mr. Umesh Aggarwal, Vice Chairman; and Mr. Shrawan Mittal, Auditor; supervised by Prof. Virendra K. Srivastava and the Convener, Prof. Sachin Jasuja did a wonderful job in coordinating various activities.

It is our duty to acknowledge the inspiring guidance of Conference General Chair, Prof. Lalit M. Patnaik, Adjunct Professor, National Institute of Advanced Studies, Indian Institute of Science, Bangalore, India, in maintaining the high standards of the conference.

We would also like to thank the student volunteers for their unforgettable, tireless, and diligent efforts.

We hope the proceedings will inspire more research in computation, communication technologies particularly in computational intelligence and metaheuristic algorithms, in the near future.

Panipat, India                                                     Ramesh K. Choudhary
Kalyani, India                                                     Jyotsna Kumar Mandal
Kolkata, India                                                 Dhananjay Bhattacharyya
December 2016

# Conference Organization

## Conference General Chair

**Lalit M. Patnaik,**
*INSA Senior Scientist and Adjunct Professor, Consciousness Studies Program,*
*National Institute of Advanced Studies, IISc Campus, Bangalore, 560012 India*

## Advisory Committee

Nikhil Ranjan Pal, Indian Statistical Institute, Calcutta, India
Pradeep K. Sinha, International Institute of Information Technology, Naya Raipur, India
Smt. Smriti Dagur, Institution of Electronics and Telecommunication Engineers, New Delhi, India
Subir Sarkar, Jadavpur University, Calcutta, India
H.A. Nagarajaram, Centre of Computational Biology, CDFD Hyderabad, India
N. Seetharamakrishna, Intel India Pvt. Ltd. Bangalore, India
P.N. Vinaychandran, Indian Institute of Science, Bangalore, India
Anthony Atkins, Staffordshire University, UK
Adrian Low, Staffordshire University, UK
Subhansu Bandopadhyay, University of Calcutta, Calcutta, India
KTV Reddy, Institution of Electronics and Telecommunication Engineers, India
Raees Ahmed Khan, B.R. Ambedkar University, Lucknow, India
Manish Bali, NVIDIA Graphics Pvt. Ltd., Bangalore, India
Cai Wen, Guandong University of Technology, China
Yang Chunyan, Chinese Association of AI, China
Li Xingsen, Nigbo Institute of Technology, China
Adrian Olaru, University Politechnica of Baucharest, Romania
Nicole Pop, UniversitateaTehnica, Romania

Bijay Baran Pal, Kalyani University, Kalyani, India
Brahmjeet Singh, National Institute of Technology, Kurukshetra, India
C. Rosen, University of Derby, Derby, UK
Chandreyee Chowdhury, Jadavpur University, Calcutta, India
Cheki Dorji, Royal University of Bhutan, Bhutan
Chowdhury Mofizur Rahman, United International University, Dhaka
D.K. Pratihar, Indian Institute of Technology, Kharagpur, India
D. Garg, Thapar University, Patiala, India
D.D. Sinha, University of Kolkata, Kolkata, India
Debasish Jana, TEOCO Software Pvt. Limited, Kolkata, India
Deepanwita Das, National Institute of Technology, Durgapur, India
Dhananjay Kulkarni, APIIT, Sri Lanka
Ingrid Rugge, University of Bremen, Germany
J.P. Choudhury, Kalyani Government Engineering College, India
Jaya Sil, Indian Institute of Engineering Science and Technology, Shibpur, India
Justin Champion, Staffordshire University, UK
K. Dasgupta, Kalyani Government Engineering College, India
K.R. Parpasani, M.A. National Institute of Technology, Bhopal, India
Kalyani Mali, University of Kalyani, Kalyani, India
Kanwalvir Singh Dhindsa, B.B.S.B. Engineering College, Punjab, India
Krishnendu Mukhopadhyay, Indian Statistical Institute, Calcutta, India
M. Sandirigam, University of Peradeniya, Peradeniya, Sri Lanka
M.K. Bhowmik, Tripura University, Agartala, India
M.K. Naskar, Jadavpur University, Kolkata, India
Mahavir Jhawar, Ashoka University, Murthal, India
Matteo Savino, University of Sannio, Italy
Md.U. Bokhari, Aligarh Muslim University, Aligarh, India
Meenakshi'Souza, Indian Institute of Information Technology, Bangalore, India
N.R. Manna, North Bengal University, Siliguri, India
Nitin Auluck, Indian Institute of Technology Ropar, Rupnagar, India
P. Jha, Indian School of Mines, Dhanbad, India
P.K. Jana, North Bengal University, Siliguri, India
P.P. Sarkar, Purbanchal University, Koshi, Nepal
Paramartha Dutta, Visvabharati University, Shantiniketan, India
Partha P. Bhattacharya, Mody University, Sikar, India
Partha S. Mandal, Indian Institute of Technology, Guwahati, India
Pitipong Yodmongkon, Chiang Mai University, Thailand
R.K. Jena, Institute of Management Technology, Nagpur, India
R.K. Samanta, North Bengal University, Siliguri, India
Rajendra Sahu, IIIT, Hyderabad, India
Rajib K. Das, University of Calcutta, Kolkata, India
Rameshwar Rijal, Kantipur Engineering College, Nepal
Rohit Kamal Chatterjee, BITMesra, Kolkata Extension Centre, Kolkata, India
S. Dutta, B.C. Roy Engineering College, Durgapur, India
S. Mal, Kalyani Government Engineering College, India

S. Mukherjee, Burdwan University, Burdwan, India
S. Muttoo, Delhi University, Delhi, India
S. Shakya, Tribhuvan University, Nepal
S.K. Mondal, Kalyani Government Engineering College, India
Sarmishtha Neogy, Jadavpur University, Kolkata, India
Sergei Silvestrov, Mälardalen University, Sweden
Shankar Duraikannan, APIIT Kualalumpur, Malaysia
Shuang Cang, Bournemouth University, UK
Siddhartha Bhattachryya, RCC Institute of Information Technology, Calcutta, India
Srabani Mukhopadhyaya, BIT Kolkata Campus, Kolkata, India
Subhamoy Changder, National Institute of Technology, Durgapur, India
Sudip Roy, Shell Technology Centre, Bangalore, India
Surya P. Singh, Indian Institute of Technology, New Delhi, India
Szabo Zoltan, Corvinus University of Budapest, Hungary
Tandra Pal, National Institute of Technology, Durgapur, India
Teresa Goncalves, Universidade de Évora, Portugal
Trupti R. Lenka, National Institute of Technology, Silchar, India
Ujjal Maulik, Jadavpur University, Calcutta, India
Utpal Sharma, Tezpur University, Tezpur, India
Vinay G. Vaidya, KPIT Technologies Ltd., Pune, India
Virendra K. Srivastava, Asia Pacific Institute of Information Technology, Panipat, India
Yacine Ouzrout, Université Lumiere Lyon2, France

## Reviewers

Ajanta Das, BIT Mesra Kolkata Extension Centre, Kolkata, India
Akash K. Tayal, I G Delhi Technical University for Women, New Delhi, India
Alok K. Rastogi, Institute for Excellence in Higher Education, Bhopal, India
Ambar Dutta, BIT Mesra Kolkata Extension Centre, Kolkata, India
Animesh Ray, Keck Graduate Institute, California, US
B.K. Tripathy, Vellore Institute of Technology, Vellore, India
Bijay Baran Pal, University of Kalyani, Kalyani, India
Chandreyee Chowdhury, Jadavpur University, Calcutta, India
D.K. Pratihar, Indian Institute of Technology, Kharagpur, India
Debasish Jana, TEOCO Software Pvt. Limited, Kolkata, India
Deepanwita Das, National Institute of Technology, Durgapur, India
Dhananjay Bhattacharyya, Saha Institute of Nuclear Physics, Calcutta, India
Himadri Sekhar Dutta, Kalyani Government Engineering College, Kalyani, India
Justin Champion, Staffordshire University, UK
Jyotsna K. Mandal, University of Kalyani, Kalyani, India
Krishnendu Mukhopadhyaya, Indian Statistical Institute, Calcutta, India

Madhurima Chattopadhyay, Heritage Institute of Technology, Calcutta, India
Mahavir Jhawar, Ashoka University, Sonepat, India
Nitin Auluck, Indian Institute of Technology Ropar, Rupnagar, India
Paramartha Dutta, Visvabharati University; Shantiniketan, India
Partha P. Bhattcharya, College of Engineering and Technology, Mody University, Sikar, India
Partha Sarathi Mandal, Indian Institute of Technology, Guwahati, India
Pradosh K. Roy, Asia Pacific Institute of Information Technology, Panipat, India
R.K. Amit, Indian Institute of Technology, Madras, India
R.R.K. Sharma, Indian Institute of Technology, Kanpur, India
Rajib K. Das, University of Calcutta, Kolkata, India
Rohit Kamal Chatterjee, BIT Mesra Kolkata Extension Centre, Kolkata, India
Sarbani Roy, Jadavpur University, Calcutta, India
Sarmishtha Neogy, Jadavpur University, Calcutta, India
Siddhartha Bhattachryya, RCC Institute of Information Technology, Calcutta, India
Somnath Mukherjee, Calcutta Business School, Calcutta, India
Srabani Mukhopadhyaya, BIT Mesra Kolkata Extension Centre, Calcutta, India
Subhamoy Changder, National Institute of Technology, Durgapur, India
Subho Chaudhuri, BIT Mesra Kolkata Extension Centre, Kolkata, India
Subir Sarkar, Jadavpur University, Calcutta, India
Sumer Singh, Indian Institute of Technology, Delhi, India
Sunandan Sen, University of Calcutta, India
Surya P. Singh, Indian Institute of Technology, New Delhi, India
Tandra Pal, National Institute of Technology, Durgapur, India
Ujjal Maulik, Jadavpur University, Calcutta, India
Utpal Biswas, Kalyani University, Kalyani, India
Utpal Sharma, Tezpur University, Tezpur

# Organizing Committee

**Patrons**

**Vinod Gupta**
Chairman, Asia Pacific Institute of Information Technology, Panipat

**Umesh Aggarwal**
Vice Chairman, Asia Pacific Institute of Information Technology, Panipat

**Shrawan Mittal**
Aiditor, Asia Pacific Institute of Information Technology, Panipat

**Convener**
Sachin Jasuja, Department of Mechatronics and EE Engineering, APIIT, Panipat

**Session Coordinators**
Arun K. Choudhary, Department of Computer Science and Engineering, APIIT, Panipat
Geeta Nagpal, Department of Computer Science and Engineering, APIIT, Panipat
Gaurav Gambhir, Department of Computer Science and Engineering, APIIT, Panipat
Shipra Chaudhuri, Department of Management, APIIT, Panipat
Pradeep Singla, Department of Electronics Engineering, APIIT, Panipat
Mahima Goel, Department of Electronics Engineering, APIIT, Panipat
Parveen Kumar, Sanjeev Sharma (Technical Support)

**Sponsorship and Finance**
Shipra Chaudhuri, Department of Management, APIIT, Panipat
Sanjeev Jawa, Department of Finance and Accounting, APIIT, Panipat

**Website Management**
Sachin Jain, Multi Media Management Unit, APIIT, Panipat

**Registration**
Rajesh Tiwari, Department of Management, APIIT, Panipat
Virender Mehla, Department of Electronics Engineering, APIIT, Panipat

Geetanjali, Librarian, APIIT, Panipat
Priyanka Sachdeva, Department of Computer Science and Engineering, APIIT, Panipat

**Hospitality, Transport, Accommodation**
Pardeep Singla, Department of Electronics Engineering, APIIT, Panipat
Afzal Khan, Administrative Officer, APIIT, Panipat

**Inventory in Charge**
Arun Choudhary, Department of Computer Science and Engineering, APIIT, Panipat
Rajesh Tiwari, Department of Management, APIIT, Panipat
Rajender Khurana, Material Management Unit, APIIT, Panipat

# Message from the General Chair

The 10th International Conference on Advanced Computing and Communication Technologies (10th ICACCT 2016) organized by the Asia Pacific Institute of Information Technology (APIIT) at Panipat during November 18–20, 2016 was primarily aimed at promoting early research on a wide range of topics in computing systems, communication technologies, and facilities. Theories of computational intelligence, it may be recalled, are being applied in many fields of science and engineering and have many other practical implementations. Therefore, the conference also aimed to enable collaborative efforts among researchers with different expertise and backgrounds.

The accepted papers were subjected to pre- and post-presentation plagiarism check and technical review. To ensure that the similarity percentage is consistently maintained within permissible limits, final versions of papers to appear in the proceedings volume were again subjected to similarity index check by Springer independently.

The Editorial Board, Advisory and Program Committee members, reviewers, invited speakers, session chairs, session coordinators, and student volunteers have done their jobs excellently towards the success of the technical program.

As General Chair of Conference, I convey my profuse thanks to Prof. Virendra K. Srivastava for providing all the excellent support from the APIIT SD India Management, which has been very generous in strongly encouraging this Conference. I would like to thank Prof. Nikhil R. Pal, Indian Statistical Institute, Calcutta; Prof. Jyotsna K. Mandal, Kalyani University; Prof. Dhananjay Bhattacharyya, Saha Institute of Nuclear Physics; Prof. Subir Sarkar, Jadavpur University; Prof. Partha P. Bhattacharya, Mody University, Sikar; Prof. Alok K. Rastogi, Institute for Excellence in Higher Education, Bhopal; Dr. Srabani Mukhopadhyay, BIT Mesra, Kolkata Extension Centre; Prof. Sarmistha Neogy, Jadavpur University; Prof. Siddhartha Bhattacharya, RCC Institute of Information Technology, Calcutta; Dr. Nitin Auluck, IIT Ropar, Rupnagar; Dr. Surya P. Singh, IIT Delhi; Prof. Atul Sharma, Thapar University, Patiala; Prof. Marjit M. Singh, NERIST, Itanagar for their valuable time and efforts in organizing the conference successfully. Finally, my special appreciation goes to Mr. Aninda Bose, Senior Editor, Springer India Pvt. Ltd., New Delhi for his advice.

We are grateful to the Institution of Electronics and Telecommunication Engineers for being the technical sponsors and Panipat Refinery and Petrochemicals, Indian Oil Corporation Ltd., Panipat, for partial financial support.

I am sure this volume will be indispensable for the researchers engaged in computing and communication technologies.

Bangalore, India                                                                Lalit M. Patnaik
December 2016                                       INSA Senior Scientist and Adjunct Professor
                                                     Conference General Chair, 10th ICACCT 2016
                                                              Consciousness Studies Program
                                                                           www.lmpatnaik.in

# Contents

# About the Editors

**Ramesh K. Choudhary**, former Director, Asia Pacific Institute of Information Technology, Panipat, obtained Ph.D. from VMU, Salem, has co-authored 'Testing of Fault Tolerance Techniques' published by Lambert Academic Publishing, Germany. He was the Indian Coordinator, Featured EUrope and South Asia MObility Network (FUSION), an ERUSMUS MUNDUS (EU) project to foster partnerships of emerging Asian countries with the EU countries reinforcing the existing collaborations developed through the EU funded projects. He had also significantly contributed to the EU-ASIA mutual recognition of studies in engineering, management, and informatics. He is Fellow of Institution of Engineers (India), IETE, and Members of IEEE, ACM, and Chinese Association for Artificial Intelligence.

**Jyotsna Kumar Mandal**, former Dean, Faculty of Engineering, Technology and Management, is Senior Professor, Department of Computer Science and Engineering, University of Kalyani, India. He has obtained Ph.D. (Engg.) from Jadavpur University. Professor Mandal has co-authored six books, viz., Algorithmic Design of Compression Schemes and Correction Techniques—A Practical Approach; Symmetric Encryption—Algorithm, Analysis and Applications: Low Cost-based Security; Steganographic Techniques and Application in Document Authentication—An Algorithmic Approach; Optimization-based Filtering of Random Valued Impulses—An Algorithmic Approach; and Artificial Neural Network Guided Secured Communication Techniques: A Practical Approach; all published by Lambert Academic Publishing, Germany. He has authored more than 350 papers on a wide range of topics in international journals and proceedings. His profile is included in the 31st edition of Marque's World Who's Who published in 2013. His areas of research include coding theory, data and network security, remote sensing and GIS-based applications, data compression, error correction, visual cryptography and steganography, distributed and shared memory parallel programming. He is Fellow of Institution of Electronics and Telecommunication Engineers, and Members of IEEE, ACM, and Computer Society of India.

**Dhananjay Bhattacharyya**, Head of Computer Sciences Division, Saha Institute of Nuclear Physics, Calcutta, India, obtained Ph.D. from Indian Institute of Science, Bangalore, and is a postdoctoral fellow from National Institutes of Health, Bethseda, Maryland, USA. Dr. Bhattacharyya's research interest is in understanding structure–function relationship of biological macromolecules, particularly nucleic acids. In this context, he utilizes available crystallographic data to carry out molecular dynamics simulation and ab initio quantum chemical calculations to understand structural stabilization of nucleic acid bases and recognition of nucleic acid by other ligands. In order to understand different structural features of nucleic acids, the CS division at Saha Institute of Nuclear Physics has developed few software tools under his guidance, such as NUPARM, BPFIND, PyrHB Find, etc. Using some of these tools, the CS division has classified structures of different non-canonical base pairs appearing in RNA crystal structures. Dr. Bhattacharyya has successfully guided several doctoral candidates on computational biology and bioinformatics.

# Part I
# Advanced Computing: Computational Intelligence

# Suitability of CORDA Model for Area Coverage Problem

**Deepanwita Das and Srabani Mukhopadhyaya**

**Abstract** COopeRative Distributed Asynchronous model or CORDA model is a basic computational model in the field of robot swarm. The objective of this work is to justify popularity and suitability of the CORDA model as a basic model of computation vis-à-vis other computational models. The problem of covering a target area has been taken up as an area of focus. This chapter presents a critical review of the various solutions of the coverage problem under CORDA model.

**Keywords** Robot swarm · Coverage · CORDA

## 1 Introduction

Robots are viewed as machines or mechanical devices which imitate the behaviour of the natural entities to carry out their jobs in an effective manner. Inspired by the coordination among the biological entities, like fish, birds, ants, bees, etc., researchers have paid huge attention in developing a system consisting of a group of miniature robots, known as *Swarm of Robots*. In some applications, a group of small and simple robots is much more effective than using a large powerful one. Simplicity and miniaturization are the deciding factors behind the acceptance of a multi-robot system over a single, powerful robot. These features result in cheaper, scalable, fault-tolerant, reusable systems of multiple robots. Robots with simple hardware support make the multi-robot systems less expensive while maintaining the performance achieved by a single powerful robot. Further, these swarms may be capable of reaching locations that are sometimes impossible for a large robot to pass through. A single robot may also not be adequately equipped to deal with a

D. Das (✉)
National Institute of Technology, Durgapur, India
e-mail: deepanwita.das@it.nitdgp.ac.in

S. Mukhopadhyaya
Birla Institute of Technology, Mesra, Kolkata Extension Centre, Kolkata, India
e-mail: smukhopadhyaya@bitmesra.ac.in

© Springer Nature Singapore Pte Ltd. 2018
R.K. Choudhary et al. (eds.), *Advanced Computing and Communication Technologies*, Advances in Intelligent Systems and Computing 562, https://doi.org/10.1007/978-981-10-4603-2_1

complex job having multiple independent modules; whereas, the inherent parallelism of the job can be exploited by a swarm. Failure of a robot in the swarm does not affect the whole job. However, robots can be added or removed from the swarm as per the requirements. The basic objective is to produce large number of inexpensive robots so that they can collaboratively and effectively perform certain jobs. As the robots in a swarm are identical, it is easy to feed them with different instructions to make them reusable in various situations.

The robots in a swarm work together in coordination with each other and are used in many applications like space explorations, search and rescue of victims, inspection, terrain mapping, oceanographic mapping, crop ploughing, etc. From application point of view, it is sometimes better to have a swarm of robots which are equipped with sensors, wheels, actuators, etc. For example, in an unknown hazardous environment sending robot rescuers is a risk-free, cost-effective way instead of sending human rescuers. Robots are capable of reaching all places or can pull a victim to a safe place through coordinated movements [1].

It is now an established fact that a swarm of robots can collectively perform several difficult tasks efficiently in a coordinated way. A swarm is expected to do various jobs like forming certain geometric patterns, gathering at any specific location, flocking like birds, partitioning themselves into groups, spreading over a region, searching specific objects, etc. Some of the applications like, demining, automated sweeping, clearing narrow AC ducts, lawn mowing, chemical or radioactive spill detection and cleanup, region filling, car body painting, etc., require scanning of each accessible portion of a whole target place. In these cases, a robot swarm can be deployed over the target region to effectively scan or cover that area. The problem of coverage of a target area by a swarm of mobile robots is getting attention by the researchers for last one or two decades.

Solvability of any problem depends on the characteristics of the robots in the swarm, as well as, the models assumed. Interestingly, in addressing the problems other than coverage, like, gathering, partitioning, etc., researchers usually consider CORDA model as a basic computational model because of its simplicity which makes the robots in the swarm easily programmable and hence, less expensive.

Several multi-robot area coverage algorithms exist in the literature. Very few of them have used CORDA as the computational model. The objective of this chapter is to present a comprehensive study of the existing solutions of coverage problem under CORDA model to justify the popularity and suitability of the model.

The organization of the paper is as follows. In Sect. 2, the CORDA model is discussed in detail. Section 3 presents a discussion on the existing solutions of the coverage problem where models of computation are non-CORDA. Section 4 describes various coverage problems and their solutions under CORDA model. Finally, we conclude in Sect. 5 indicating some interesting challenges for future scope of work.

## 2   CORDA Model: A Basic Computational Model

*Cooperative Distributed Asynchronous* model or CORDA [2] is a theoretical model. It is a step towards reducing the cost of the system to emulate real-life activities. In this model, the activity of any robot can be viewed (and hence programmed) as a sequence of computational cycles when each cycle consists of four phases, namely, wait (optional), observe, compute and move. These phases are non-overlapping and each phase is carried out sequentially by a robot. In each cycle, a robot first observes its surroundings by means of taking a snapshot of its neighbourhood using the camera or sensor attached to it. In the second phase, it computes its next destination according to the observation made in the *observe* phase. It then moves to its computed destination in *move* phase. Once such a cycle is completed, the robot starts the next cycle and continues to do so until the whole job is completed.

The three phases of a computational model are executed by the robots in sequence. Thus, when a robot looks at its surroundings, it does not move or compute; when it computes, it neither moves nor looks and when it goes towards its computed destination, it does not do anything else. These features make the robots easier to program and thus make the solution cheaper. Thus, this model supports the basic objective of using swarm of robots instead of a single, powerful, expensive robot.

Interestingly, this model does not, in general, imply existence of any direct communication among the robots through message passing. On the contrary, here the only mode of communication among the robots is through observing the positions of others. A snapshot of the surrounding shows only the position of a robot but not the phase in which the robot was at the time of taking the snapshot. Thus, on the basis of a snapshot when a robot decides it next course of action, it is possible that its neighbours might have already changed their positions. Thus, synchrony among the robots is very crucial for this computational model.

In a synchronous system, all robots take snapshot of the surroundings at the same time. Hence, each of them computes its destination on the basis of same snapshot as long as their visibility range is unlimited. In case of limited visibility, the snapshots may differ depending on their visibility range but at least it is guaranteed that all robots are in the same phase. In semi-synchronous system, all robots are not always active. However, active robots are synchronous. Asynchronous system under CORDA model is the most challenging and interesting area of research.

## 3   Earlier Models

Coverage using multiple robots is classified into two categories: team-based coverage and individual coverage. Earlier works are mostly team-based. In team-based approach, the members (robots) of a team form a horizontal line by positioning themselves side by side and the whole team moves along a particular direction to cover the given area. A critical point is detected when the continuous movement of the team gets obstructed due to the presence of an obstacle. The first point of contact on the obstacle is known as the critical point with respect to that obstacle. Upon detection of a critical point, a team splits itself into sub-teams. Each of these sub-teams now covers the regions on either sides of that obstacle and they rejoin again on completion.

Based on this approach, Latimer et al. [3] proposed a multi-robot coverage algorithm. Every teammate stores and shares a common adjacency graph. The graph represents the covered and uncovered subregions, say cells, inside the target region associated with that particular team. For each occurrence of team division and rejoining that graph is updated by all the team members. When all the teams explore all the cells in their individual adjacency graphs, the procedure terminates.

Kong et al. [4] proposed an algorithm in which, the robots determine whether the cell to be covered is divided into disconnected subregions by any obstacle or not. Each robot shares an adjacency graph that represents adjacent subregions to be covered and the disconnected parts of those regions, if any. After covering a cell, a robot updates and broadcasts its graph.

Rekletis et al. [5, 6] proposed two coverage approaches: *Team-based* for restricted communication and *Individual* for unrestricted communication. In [5], the communication is restricted to the line of sight (LOS). A Team contains two types of robots: explorers and coverers. There are mainly two explorers in a team that explore the region, one along each of the top and lower boundaries of the target region. To do so, these explorers need to maintain same lateral speed and the explorers detect the presence of an obstacle inside the region when they lose their LOS. On the other hand, the *coverers* cover the free intermediate zone. In [6], the target area is partitioned into virtual strips equal to the number of robots. The robots are deployed at regular intervals each along one side of a strip to be covered and a robot moves around its strip boundary to explore it. In both the cases, the robots gather the knowledge of critical points, build global *Reeb graphs*, which are shared and updated by all the robots.

In all the cases discussed above, sharing and upgrade of graphs need full synchrony among the team members. Moreover, the members of a team simultaneously carry out multiple jobs, like covering, observing, detection of obstacles, sharing of information, communicating with team mates, etc. These features are more suitable to powerful robots than a small, simple robot usually constituting a swarm. In all these solutions, inter-robot communication is also absolutely necessary to achieve the goal. As a result, message passing overhead is high. Under the circumstances, CORDA model offers a better alternative for computation.

# 4   Coverage Under CORDA Model

The main objective of distributed coverage is to create a reliable, fault-tolerant and fully decentralized algorithm through which robots can cover a target region without repeated coverage and collision. It is desirable that the robots work in a completely distributed environment having a silent communication among them. The aim is to design the solutions considering models which are closer to reality than the previous ones. Researchers in this field usually consider CORDA model as the basic computational model because of its closer approximation to the real-life situation. The algorithms discussed in this section are based on CORDA model or wait-observe-compute-move model. Different variations of the coverage problem under CORDA model are as follows:

## A. Covering of a target area without obstacle

Let us first discuss the solutions when target area is free of obstacles. In [7], CORDA model is assumed for asynchronous robots. The robots considered here are having unlimited visibility. In this algorithm, the target area is partitioned into several blocks (subregions) by the robots themselves. The number of the blocks is equal to the number of robots located within the target region. In observe phase, robots observe each other's location and according to that they rank themselves in an independent manner. Since the robots are asynchronous, the snap shots taken are not identical in general. The ranking is based on the locations of the robots and it has been shown that all the robots unanimously agree on this relative ranking, in spite of the fact that they do not agree on the orientation of one of the axes and do not exchange any message among themselves. A robot identifies its cell according to its rank in compute phase. It then moves to that cell to start the job of coverage. A robot may stop at any position while moving towards destination. In that case it once again starts another computational cycle. Once each of the robots completely covers its own cell, the coverage is complete. The algorithm results in distributed coverage of a target area within finite amount of time without any repetition and collision.

In this solution a robot cannot successfully rank all the robots in the swarm if it has limited visibility. In reality, robots can see only a circular area of fixed radius. In [8], the authors assumed the robots to have limited visibility which is more realistic assumption than what it is in [7]. To meet the challenge of limited visibility, robots and models are made slightly powerful than [7]. The robots are assumed to follow asynchronous model with an added restriction that all robots are always active and they start execution at the same time. In this solution, the robots require to see their nearest (with respect to the horizontal distance only) left and right neighbours at least once throughout the whole process. These neighbours might not be visible to the robot initially because of its limited visibility. However, as the robots are asynchronous it is not even guaranteed that the neighbours would be visible to each other at all, especially when their velocities differ. The authors have presented a solution which is valid when the initial distribution of the robots is dense. In this solution, at first a robot calculates the boundaries of the strip that is to be covered by itself by looking at its nearest horizontal neighbours. In the next phase, the actual coverage of the strip is carried out by the robots.

## B. Covering of a target area with obstacles

A solution is presented in [9] for covering a target area with obstacles present inside the region. As the difficulty level of this problem is higher than that of the previous ones, the solution assumes slightly more powerful robots. The robots here are assumed to have unlimited visibility range. This solution is also based on CORDA as the model of computation. The proposed solution assumes that the robots are synchronous.

Initially, the robots are assumed to be distributed randomly over the whole region and then they are made to assemble on the left boundary using the approach discussed in [10]. From the left boundary, the robots start the execution. The robots partition the whole area into blocks and paint the area block-wise. Initially, a robot tries to find out if any other robot is present inside the target region below it. If none is found and its vision is obstructed by any obstacle, it moves directly to that obstacle, the entry point of the next block, to explore that block. Otherwise, it starts painting the current block, an area above the horizontal line passing through its current position. The algorithm guarantees painting of the whole target region in finite amount of time and without any repetition and collision.

## 5   Conclusion

CORDA model is a well-accepted computational model because of its suitability in the field of swarm. In this work, we have included a detailed discussion on CORDA model and have justified the use of CORDA as a computational model in reference to the problem of *multi-robot area coverage.*

Velocities of the robots play a crucial role for CORDA model under limited visibility. In the field of multi-robot area coverage, this issue has not yet been addressed so far. This could have been an interesting area of research.

## References

1. Mondada, F., et al.: Transport of a child by swarm-bots. Laboratory of Intelligent Systems, Swiss Federal Institute of Technology. Available from: www.youtube.com/watch?v= CJOubyilTsE Accessed: 01 July 2007
2. Prencipe, G.: Corda: distributed coordination of a set of autonomous mobile robots. In: 4th European Research Seminar on Advances in Distributed Systems, pp. 185–190 (2001)
3. Latimer, D., Srinivasa, S., Lee Shue, V., Sonne, S., Choset, H., Hurst, A.: Toward sensor based coverage with robot teams. In: IEEE International Conference on Robotics and Automation, vol. 1, pp. 961–967 (2002)
4. Kong, C., Peng, N.A., Rekletis, I.: Distributed coverage with multi-robot system. In: IEEE International Conference on Robotics and Automation, pp. 2423–2429 (2006)
5. Rekletis, I., Lee Shue, V., New, A.P., Choset, H.: Limited communication, multi-robot team based coverage. In: IEEE International Conference on Robotics and Automation, pp. 3462–3468 (2004)
6. Rekletis, I., New, A.P., Rankin, E.S., Choset, H.: Efficient boustrophedon multi robot coverage: an algorithmic approach. Ann. Math. Artif. Intell. **52**(2–4), 109–142 (2008)

7. Das, D., Mukhopadhyaya, S.: An algorithm for painting an area by swarm of mobile robots. Int. J. Inform. Process. **7**(3), 1–15 (2013)
8. Das, D., Mukhopadhyaya, S.: Distributed algorithm for painting by a swarm of robots under limited visibility. Int. J. Adv. Robot. Syst. 2016
9. Das, D.: Distributed area coverage by swarm of mobile robots. Doctoral Thesis, Submitted at National Institute of Technology, Durgapur, India (2016)
10. Das, D., Mukhopadhyaya, S.: Multi-robot assembling along a boundary of a given region in presence of opaque line obstacles, pp. 21–29. ICICA, Chennai (2015). doi:10.1007/978-981-10-1645-5-3

# A Hybrid of Fireworks and Harmony Search Algorithm for Multilevel Image Thresholding

Shivali, Lalit Maurya, Ekta Sharma, Prasant Mahapatra
and Amit Doegar

**Abstract** Multilevel image thresholding is an essential part of image processing. This paper presents a hybrid implementation of fireworks and harmony search algorithm where Kapur's entropy is used as the fitness function for solving the problem. The results of the proposed method have been compared with the standard fireworks algorithm (FWA) and particle swarm optimization (PSO) based multilevel thresholding methods. Experimental results indicate that the proposed method is a promising approach in the field of image segmentation.

**Keywords** Image segmentation · Multilevel image thresholding · Fireworks algorithm · Harmony search algorithm · Kapur's entropy

## 1 Introduction

Image segmentation based on multilevel thresholding technique is an intriguing and critical task. There are various techniques for image segmentation, out of which thresholding is the one largely used because of its easy implementation. Partitioning

Shivali · L. Maurya · E. Sharma · P. Mahapatra (✉)
Biomedical Instrumentation Division (V-2), CSIR-Central Scientific Instruments
Organisation, Sector-30, Chandigarh 160030, India
e-mail: prasant22@csio.res.in

Shivali
e-mail: shivali_katnoria@yahoo.com

L. Maurya
e-mail: lalitmaurya47@gmail.com

E. Sharma
e-mail: ektasharma0013@gmail.com

Shivali · E. Sharma · A. Doegar
Department of Computer Science & Engineering, National Institute of Technical Teachers'
Training and Research, Sector-26, Chandigarh 160019, India
e-mail: amit@nitttrchd.ac.in

© Springer Nature Singapore Pte Ltd. 2018
R.K. Choudhary et al. (eds.), *Advanced Computing
and Communication Technologies*, Advances in Intelligent Systems
and Computing 562, https://doi.org/10.1007/978-981-10-4603-2_2

11

the original image into various numbers of classes for extraction of meaningful information [1] is defined as image thresholding. If the image is segmented in two classes namely the foreground and the background, it is termed as bi-level thresholding. The concept can further be extended to multilevel thresholding for attaining more than two classes [2].

However, it necessitates an optimal value of threshold to separate foreground from their background. The main aim of thresholding is to select best value of threshold [1]. Till date, numerous methods for thresholding have been developed. A criterion called Otsu criteria [3] was proposed which maximizes between class variance for getting segmented regions of the image. Moment preserving approach was used by Tsai [4] for selecting threshold of grayscale images, which is referred to as Tsallis entropy. Kapur et al. [5] used entropy of the histogram for thresholding. According to Kapur's criterion, entropy of each class or sum of entropies was maximized based on information theory.

Throughout the years, many heuristic algorithms including genetic algorithm [6], particle swarm optimization [7], firefly algorithm [8], bacterial foraging algorithm [9], ant colony optimization [10], harmony search algorithm [11], cuckoo search algorithm [12] and fireworks algorithm [13] have been used for image thresholding. In this paper, to find the optimal value of threshold for image segmentation, fireworks algorithm is hybridized with harmony search algorithm and Kapur's entropy is used as fitness function.

The remaining part of the paper is structured as follows: Sect. 2 gives introduction to fireworks algorithm, Sect. 3 gives introduction to harmony search algorithm. Proposed methodology is outlined in Sect. 4. The experimental results are shown in Sect. 5. The conclusions are drawn in Sect. 6.

## 2 Fireworks Algorithm (FWA)

```
Select random n locations for fireworks
while maximum function evaluations is not reached do
Set off n fireworks at n locations
for each firework x_i do
Calculate the number of sparks S_i and amplitude for fire-
work A_i
Obtain locations of S_i sparks of firework x_i
end for
for k = 1 : number of Gaussian sparks do
Select random firework x_j
Generate a Gaussian spark for selected firework
end for
```

```
Select best firework for next iteration
Select randomly n - 1 from two types of sparks and fire-
works based on probability
```
**end while**

FWA has two types of fireworks: well-manufactured firework and lower quality firework. It has four parts: explosion operator, mutation operator, mapping strategy and selection strategy [14]. Explosion operator generates sparks around the fireworks. Mutation operator governs the number and the amplitude of sparks.

# 3 Harmony Search Algorithm (HSA)

In HSA [15], every solution is known as harmony which is denoted by $n$-dimensional real vector. First, an initial population of harmony vectors is generated arbitrarily and is kept in the harmony memory (HM). A new candidate of harmony is produced using either pitch adjustment operation or random re-initialization from the elements in HM. Finally, HM gets updated by comparison of new harmony and worst harmony. The above process keeps on repeating itself until the stopping criterion fulfills.

# 4 Proposed Methodology

In general, the multilevel thresholding based on the optimization algorithm uses the objective function which has to maximize by the optimization algorithm. In this work, Kapur's method is used to define the objective function for the thresholding and the proposed hybrid algorithm has to find out the threshold level by considering Kapur's method.

## 4.1 Kapur's Entropy Function

Kapur's entropy is used in this paper for image thresholding and is formulated as follows [13]:

Let an image $I$ contains $n$ number of pixels whose gray level lies between 0 and $L - 1$. Let $h_i$ denote the number of pixels at gray level $i$, and the likelihood of occurrence of gray level $i$ in the image is denoted by $\text{pr}_i$ as shown in Eq. (1)

$$\mathrm{pr}_i = h_i/n, \tag{1}$$

where the $k$ dimensional problem is a division of an image into $k + 1$ classes and obtaining $k$ optimal thresholds $(t_0, t_1, \ldots t_{k-1})$. The objective function as mentioned in the Eq. (2) is maximized, and hence the optimal value of threshold is obtained.

$$f(t_0, t_1 \ldots t_{k-1}) = \sum_{i=0}^{k} H_i \tag{2}$$

where, entropies $H_i$ is defined as:

$$H_0 = -\sum_{i=0}^{t_0-1} \frac{\mathrm{pr}_i}{w_0} \ln \frac{\mathrm{pr}_i}{w_0}, \quad w_0 = \sum_{i=0}^{t_0-1} \mathrm{pr}_i \tag{3}$$

$$H_1 = -\sum_{i=t_0}^{t_1-1} \frac{\mathrm{pr}_i}{w_1} \ln \frac{\mathrm{pr}_i}{w_1}, \quad w_1 = \sum_{i=t_0}^{t_1-1} \mathrm{pr}_i \tag{4}$$

$$H_k = -\sum_{i=t_{k-1}}^{L-1} \frac{\mathrm{pr}_i}{w_k} \ln \frac{\mathrm{pr}_i}{w_k}, \quad w_k = \sum_{i=t_{k-1}}^{L-1} \mathrm{pr}_i \tag{5}$$

The threshold at which function returns maximum value is considered as the optimal value of threshold.

## 4.2 Proposed Hybrid Algorithm: FWA/HSA

In standard FWA, the fireworks with better objective value generate a larger explosion sparks within range (within small explosion amplitude). Conversely, fireworks with low objective value generate a smaller explosion spark within a smaller range. These characteristics allow balancing the exploration (diversification) and exploitation (intensification). However, it has been found that FWA shows a premature convergence for the function which does not have their optimum at the origin. In this work, the standard FWA convergence behavior is improved by introducing harmony search algorithm for the multilevel thresholding problem. The harmony search algorithm provides a better balance of intensification and diversification. In proposed hybrid method diversification is better controlled by the pitch adjustment and random selection.

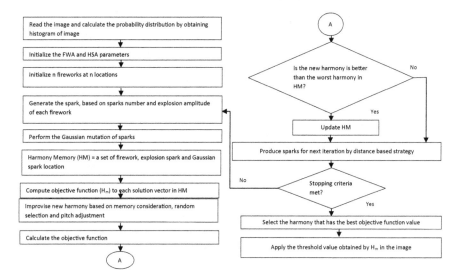

**Fig. 1** Flow chart of proposed algorithm

In the proposed hybrid method, at each iteration FWA produces a set (called as seedmatrix) of firework, explosion spark and Gaussian spark location. The HM corresponds to the each particle generated by FWA, i.e., all solutions of FWA stored in the HM. After that the harmony search operation, namely memory consideration, random selection and pitch adjustment operator are performed on the set of firework, explosion spark and Gaussian spark location. If the new harmony obtained after the harmony operation is better than the worst harmony, it is replaced by the worst harmony in HM. Therefore, introduction of harmony search operation improves the diversity of the standard FWA. It increases the speed of global convergence and reduces the chances of stuck in local optima. At each iteration, the distance-based strategy is used to select the best spark from seedmatrix as a firework of the next iteration. The flow chart of the proposed hybrid algorithm for multilevel thresholding has been shown in Fig. 1.

# 5 Experimental Results

The proposed hybrid algorithm FWA/HSA has been applied to some standard test images and lathe tool's images. The reason for using lathe tool's images is that the authors used the tool to carry out different experiments on micro and nano scale movements for desktop machining [16]. The proposed method has been tested on

**Table 1** Simulation results
of proposed hybrid algorithm

| Image | $K^{a}$ | Threshold values | Objective function |
|---|---|---|---|
| Pirate | 1 | 102 | 13.0784 |
|  | 2 | 89, 170 | 18.1847 |
|  | 3 | 59, 114, 171 | 22.7723 |
|  | 4 | 48, 89, 131, 173 | 26.9231 |
| Clock | 1 | 134 | 12.7189 |
|  | 2 | 32, 131 | 17.7190 |
|  | 3 | 32, 102, 168 | 22.5607 |
|  | 4 | 32, 91, 144, 197 | 27.0509 |
| Elaine | 1 | 151 | 12.9879 |
|  | 2 | 107, 175 | 17.9454 |
|  | 3 | 91, 143, 193 | 22.3877 |
|  | 4 | 22, 93, 148, 198 | 26.8252 |
| Home | 1 | 117 | 13.3231 |
|  | 2 | 83, 168 | 18.5773 |
|  | 3 | 65, 125, 186 | 23.2543 |
|  | 4 | 55, 107, 159, 204 | 27.5938 |
| Ref | 1 | 23 | 9.4304 |
|  | 2 | 20, 82 | 14.1644 |
|  | 3 | 26, 145, 166 | 19.6226 |
|  | 4 | 20, 83, 145, 166 | 24.2756 |
| Move 2mm | 1 | 21 | 9.4042 |
|  | 2 | 20, 84 | 14.2457 |
|  | 3 | 27, 146, 166 | 19.6351 |
|  | 4 | 19, 82, 146, 165 | 24.2693 |
| Move 4mm | 1 | 24 | 9.3409 |
|  | 2 | 21, 82 | 14.1449 |
|  | 3 | 23, 144, 164 | 19.5968 |
|  | 4 | 24, 144, 164, 209 | 24.0126 |

[a]$K$ is the level of segmentation

all the images available in [17] but here only four images have been shown. The parameters set for hybrid algorithm are number of fireworks—5, highest value of explosion amplitude Â—40, Spark number coefficient ($m$)—50, HMCR—0.95, PAR—0.5, BW—0.5. The stopping criterion is selected as a number of iteration in which best fitness remains constant at 10% of the number of iterations (NI) or max number of iterations is achieved. Table 1 shows the results of the proposed hybrid algorithm. Figure 2 shows the segmented result of selected images with

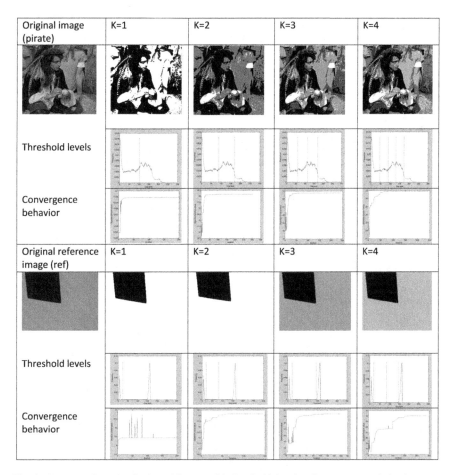

**Fig. 2** Segmented result of selected image with threshold level and convergence behavior

convergence behavior and threshold level. In Table 2, the proposed hybrid algorithm has been compared with the PSO and standard FWA based multilevel thresholding. The algorithm is run 25 times for each value of k over each image. For each image, mean objective function value, standard deviation, and computation time are calculated. Table 2 shows that the hybrid algorithm based method's mean values of objective functions over 25 runs are superior to other methods. The computational time of proposed hybrid FWA/HSA based multilevel thresholding method is less than PSO and standard FWA based method because it shows better and fast convergence behavior and mostly converge before the max number of iterations. It has been observed that the proposed algorithm is superior to the other algorithm and give excellent solution. It has also been observed that the proposed algorithm gives superior result in terms of computational time because it shows the fast convergence behavior.

**Table 2** Standard deviation and computation time of PSO, standard FWA and proposed method

| Image | $K$ | PSO | | | FWA | | | Proposed hybrid FWA/HSA | | |
|---|---|---|---|---|---|---|---|---|---|---|
| | | Mean | Standard deviation | Time (s) | Mean | Standard deviation | Time (s) | Mean | Standard deviation | Time (s) |
| Pirate | 2 | 13.0783 | 1.2796e−04 | 10.0503 | 18.060 | 0.6205 | 10.0882 | 18.0528 | 7.0030e−05 | 3.2112 |
| | 3 | 18.1498 | 0.0385 | 14.5952 | 21.3125 | 1.0206 | 10.9682 | 21.4766 | 0.0739 | 8.0042 |
| | 4 | 22.5628 | 0.2023 | 11.2853 | 25.7642 | 1.3337 | 14.0045 | 26.2171 | 0.0443 | 7.5213 |
| | 5 | 26.1371 | 0.4543 | 14.8767 | 29.9721 | 1.5348 | 14.8596 | 30.4197 | 0.2130 | 11.1767 |
| Clock | 2 | 12.7189 | 3.1171e−04 | 10.4248 | 17.3671 | 0.8162 | 12.1728 | 17.2433 | 0.0741 | 4.2243 |
| | 3 | 17.5770 | 0.0947 | 14.0783 | 21.2537 | 1.1931 | 12.0240 | 21.2702 | 0.2073 | 9.3374 |
| | 4 | 22.0902 | 0.6433 | 15.1192 | 25.3670 | 0.9401 | 13.1073 | 25.5546 | 0.6213 | 11.1652 |
| | 5 | 25.5236 | 1.6407 | 15.7067 | 29.3215 | 1.1934 | 13.3215 | 29.7579 | 1.0752 | 9.1358 |
| Elaine | 2 | 15.3425 | 0.0078 | 10.4651 | 17.9453 | 4.3660e−05 | 10.7238 | 17.8265 | 0.0039 | 4.7315 |
| | 3 | 17.8932 | 0.5821 | 14.8290 | 21.2731 | 0.9585 | 11.2731 | 21.7786 | 0.9122 | 6.5900 |
| | 4 | 23.5471 | 1.0452 | 14.0281 | 25.9743 | 0.3778 | 11.9915 | 25.9511 | 0.4017 | 8.4311 |
| | 5 | 25.5047 | 1.4024 | 14.7398 | 30.6002 | 0.8375 | 12.5938 | 30.4646 | 0.9885 | 7.9989 |
| Home | 2 | 13.3108 | 0.4714 | 10.8322 | 18.4387 | 0.6866 | 10.1472 | 18.4358 | 0.6768 | 8.9142 |
| | 3 | 18.4232 | 1.3221 | 11.2887 | 21.9726 | 1.1585 | 10.9291 | 21.9617 | 1.1493 | 9.4964 |
| | 4 | 22.5021 | 1.2023 | 14.0121 | 26.5845 | 0.8698 | 11.7494 | 26.5678 | 0.9389 | 8.5269 |
| | 5 | 26.2171 | 1.9231 | 14.8943 | 30.9394 | 1.7066 | 12.3634 | 30.9736 | 1.6124 | 7.6879 |
| Reference | 2 | 13.5671 | 0.6781 | 10.0361 | 14.5422 | 0.514102 | 10.3304 | 14.4477 | 0.4671 | 5.1242 |
| | 3 | 17.1292 | 1.8391 | 15.0283 | 18.7125 | 1.2052 | 11.2437 | 18.5380 | 1.3870 | 7.6134 |
| | 4 | 20.0831 | 1.4568 | 14.0289 | 22.0982 | 1.5032 | 11.8846 | 22.0189 | 1.4425 | 10.2289 |
| | 5 | 23.8093 | 1.9563 | 14.9210 | 25.9076 | 1.5498 | 12.9770 | 25.3118 | 1.0948 | 11.0628 |

(continued)

**Table 2** (continued)

| Image | K | PSO | | | FWA | | | Proposed hybrid FWA/HSA | | |
|---|---|---|---|---|---|---|---|---|---|---|
| | | Mean | Standard deviation | Time (s) | Mean | Standard deviation | Time (s) | Mean | Standard deviation | Time (s) |
| Move 2 mm | 2 | 13.0567 | 0.5309 | 9.9312 | 14.3621 | 0.4489 | 10.3205 | 14.8227 | 0.5231 | 5.4890 |
| | 3 | 17.2034 | 1.0382 | 10.9283 | 18.4290 | 1.6082 | 12.3891 | 18.3221 | 1.6455 | 9.4725 |
| | 4 | 20.1139 | 1.2389 | 14.8292 | 22.1380 | 1.5138 | 13.6345 | 22.4343 | 1.4345 | 10.8052 |
| | 5 | 23.8930 | 1.5930 | 14.9723 | 26.0206 | 1.9142 | 14.6207 | 25.0626 | 1.3318 | 14.2404 |
| Move 4 mm | 2 | 13.0562 | 0.4821 | 10.0231 | 14.5530 | 0.5797 | 12.7458 | 14.3539 | 0.4824 | 5.9210 |
| | 3 | 17.2730 | 1.3820 | 14.8293 | 18.1527 | 1.8324 | 13.5795 | 18.5650 | 1.2327 | 8.4001 |
| | 4 | 20.4511 | 0.9322 | 14.7282 | 22.4778 | 1.9145 | 14.5243 | 21.9931 | 1.3823 | 13.4602 |
| | 5 | 23.8948 | 1.5920 | 14.9212 | 25.3100 | 1.2854 | 13.7559 | 25.4694 | 1.2320 | 12.1681 |

# 6   Conclusion

In this paper, a hybrid of standard fireworks algorithm (FWA) and harmony search algorithm (HSA) based multilevel thresholding has been proposed. In proposed hybrid algorithm, the fireworks algorithm has been improved by introducing harmony search operations, viz., memory consideration, random selection, and pitch adjustment operator. Introduction of harmony search operation improves the diversity of the standard FWA. It increases the speed of global convergence and reduces chances of getting trapped in local optima. In the standard FWA, the initial best value is selected for next iteration and remaining $n - 1$ values are selected on the basis of distance-based strategy. In the proposed technique, harmony search operations are performed on the matrix which contains initial fireworks, explosion sparks, Gaussian sparks. Then, old matrix is compared to the new one and the best values among both are kept in the new matrix. Then, best firework from the new matrix is selected to be passed to the next iteration and remaining $n - 1$ fireworks are selected as per the procedure of the standard FWA. This improves the time complexity (optimum value is reached in less time) and also reduces the chance to get trapped in local optima solutions.

Particle swarm optimization (PSO) has been widely used for image thresholding recently, hence, the proposed algorithm has been compared to PSO and in that case also the proposed method is found to be superior.

**Acknowledgements**   This work is supported by the Council of Scientific and Industrial Research (CSIR, India), New Delhi under the Network program (ESC-112) in collaboration with CSIR–CMERI, Durgapur. Authors would like to thank Director, CSIR–CSIO for his guidance during the investigation.

# References

1. Kumar, A., Kumar, V., Kumar, A., Kumar, G.: Expert systems with applications cuckoo search algorithm and wind driven optimization based study of satellite image segmentation for multilevel thresholding using Kapur's entropy. Expert Syst. Appl. **41**, 3538–3560 (2014)
2. Nabizadeh, S., Faez, K., Tavassoli, S., Rezvanian, A.: A novel method for multi-level image thresholding using particle swarm optimization algorithms. IEEE Int. Conf. Comput. Eng. Technol. **4**, V4271–V4275 (2010)
3. Otsu, N.: A threshold selection method from gray-level histograms. IEEE Trans. Syst. Man Cybern. **9**(1), 62–66 (1979)
4. Tasi, W.: NOTE moment-preserving thresholding. In: Document Image Analysis. IEEE Computer Society Press (1985)
5. Kapur, J.N., Sahoo, P.K., Wong, A.K.C.: A new method for gray-level picture thresholding using the entropy of the histogram. Comput. Vis. Graph. Image Proc. **29**(3), 273–285 (1985)
6. Zhao, X., Lee, M., Kim, S.: Improved image segmentation method based on optimized threshold using genetic algorithm. **2**, 921–922 (2008)
7. Wei, C., Kangling, F.: Multilevel thresholding algorithm based on particle swarm optimization for image segmentation. In: 27th Chinese Control Conference. IEEE, pp. 0–3 (2008)

8. Horng, M.-H., Jiang, T.-W.: Multilevel image thresholding selection using the artificial bee colony algorithm. In: International Conference on Artificial Intelligence and Computational Intelligence, vol. 6320, pp. 318–325. Springer, Berlin, (2010)
9. Sathya, P.D., Kayalvizhi, R.: Image segmentation using minimum cross entropy and bacterial foraging optimization algorithm. In: International Conference of Emerging Trends in Electrical and Computer Technology (ICETECT), pp. 500–506 (2011)
10. Zhao, X., Lee, M.-E., Kim, H.-S.: Improved image thresholding using ant colony optimization algorithm. In: International Conference of Advanced Language Processing and Web Information Technology. IEEE, pp. 210–215 (2008)
11. Alia, O.M., Mandava, R., Ramachandram, D., Aziz, M.E: A novel image segmentation algorithm based on harmony fuzzy search algorithm. In: International Conference of Soft Computing and Pattern Recognition. IEEE, pp. 335–340 (2009)
12. Zhao, W., Ye, Z., Wang, M., Liu, W: An image threholding approach based on cuckoo search algorithm and 2D maximum entropy. In: 8th International Conference on Intelligent Data Acquisition and Advanced Computing Systems: Technology and Applications (IDAACS). IEEE, no. 2014207, pp. 303–308 (2015)
13. Tuba, M., Bacanin, N., Alihodzic, A.: Multilevel image thresholding by fireworks algorithm. In: 25th International Conference on Radioelektronika (RADIOELEKTRONIKA). IEEE, pp. 326–330 (2015)
14. Tan, Y.: Fireworks algorithm: a novel swarm intelligence optimization method, 1st ed., p. 323. Springer Publishing Company, Incorporated, (2015)
15. Oliva, D., Cuevas, E., Pajares, G., Zaldivar, D. Perez-Cisneros, M.: Multilevel thresholding segmentation based on harmony search optimization. J. Appl. Math. (2013)
16. Mahapatra, P.K., Thareja, R., Kaur, M., Kumar, A: A machine vision system for tool positioning and its verification. Meas. Control 48(8), 249–60 (2015)
17. http://sipi.usc.edu/databaseaccessed. Accessed on 01 Feb 2016

# Performance Evaluation of New Text Mining Method Based on GA and *K*-Means Clustering Algorithm

**Neha Garg and R.K. Gupta**

**Abstract** Rapid breakthrough in technology and reduced storage cost permit the individuals and organizations to generate and gather an enormous amount of text data. Extracting user interested documents from this gigantic amount of text data is a tedious job. This necessitates the development of text mining method for discovering interesting information or knowledge from the massive data. Document clustering is an effective text mining method which classifies the similar set of documents into the most relevant groups. *K*-means is the most classic clustering algorithm. However, results obtained by *K*-means highly depend on initial cluster centers and might be trapped in local optima. The paper presents a *K*-means document clustering algorithm with optimized initial cluster centers based on genetic algorithm. Experimental studies conducted over two different text datasets confirm that clustering results are more accurate by the application of the proposed method compared to *K*-means clustering.

**Keywords** Document clustering · Genetic algorithm · *K*-means · Purity · *F*-measure

## 1 Introduction

In today's world, the volume of text data is mounting rapidly which is available in various types such as web, news sites, government administrations, organization wide intranets, extranets, etc. As there is gigantic amount of text data on the web and other repositories, there is a requirement for the development and use of data analysis tools and techniques to locate the relevant information or knowledge amongst these massive collections. So, to achieve the task of organizing such

N. Garg (✉) · R.K. Gupta
Department of CSE & IT, Madhav Institute of Technology and Science, Gwalior, India
e-mail: nehagarg179@gmail.com

R.K. Gupta
e-mail: iiitmrkg@gmail.com

© Springer Nature Singapore Pte Ltd. 2018       23
R.K. Choudhary et al. (eds.), *Advanced Computing*
*and Communication Technologies*, Advances in Intelligent Systems
and Computing 562, https://doi.org/10.1007/978-981-10-4603-2_3

massive collections, text mining comes into play as it provides very useful tool for clustering.

Clustering is a useful text mining method that subdivides the documents into desired number of clusters, so that documents of same group have higher similarity and documents of different groups will have large differences [1]. The $K$-means [2] is the most classic clustering algorithm that arranges the documents in order such that the similar documents are located close to each other. However, in $K$-means user have to define the appropriate number of clusters and their initial clusters centers in advance, which may some of the time hard to set [3].

In recent years, a number of methods have found that use the $K$-means algorithm [4] for overcoming its shortcomings. Steinbach et al. [5] present comparative results of agglomerative hierarchical clustering with standard and bisecting $K$-means for document clustering. Their performances are evaluated by overall similarity, $F$-measures and entropy. Final results prove that the bisecting $K$-means performs better than hierarchical and standard $K$-means method.

A new method was proposed by Sihag et al. [6] for the improvement in $K$-means algorithm. They introduced a graph-based method which computes the centrality of each node using dissimilarity and cohesiveness value and those nodes having high centrality value are taken as initial clusters centers. They find that graph-based method gives a significant improvement over $K$-means method.

Premalatha and Natarajan [7] defines a genetic algorithm along with simultaneous and ranked mutation for enhancing the $K$-means performance in clustering of text documents. Simultaneous mutation defines that GA at once uses various mutation operators for producing next generation. However, rank mutation defines that mutation depends on chromosomes fitness rank of previous population.

This article presents a $K$-means clustering algorithm based on genetic algorithm (GA). In general, GA is the commonly used evolutionary algorithm that conducts a globalized searching for optimal clustering. The $K$-means algorithm tends to converge faster than the genetic algorithm, but may get trapped in local optima [8] because its performance highly depends on the initial clusters' centroids. Therefore, for improving the performance, the hybrid algorithm is proposed.

The paper is organized as follows: The proposed method for clustering of documents is described in Sect. 2. Section 3 presents the performance evaluation by comparing the results of proposed algorithm and $K$-means algorithm. Section 4 concludes the paper.

## 2 Methodology

Document clustering is a challenging task in text mining because of the high-dimensional text documents. Therefore, it requires effective as well as efficient clustering algorithm which can address this issue. To apply any clustering algorithms on text corpus, it is important to produce document vector from dataset, with the goal that documents are represented in an appropriate form for clustering.

## 2.1 Document Representation

There are different ways to represent the documents, the most common representation comprise the Vector Space Model (VSM). In VSM, the statistics of terms present in documents are characterized by some measures as term frequency (tf) that defines the number of times a term is present in the document and inverse document frequency (idf) which defines the occurrence of term in different documents [9]. The inverse document frequency $idf_j$ for term $t_j$ is measured as:

$$idf_j = \log \frac{N}{df_j} \tag{1}$$

where $N$ represents the whole amount of documents in the dataset and $df_j$ denotes how many documents have term $t_j$ present.

Now, the term weight of $t_j$ in document $d_i$ will be:

$$wt_{ij} = tf_{ij} * idf_j \tag{2}$$

Therefore, the document $d_i$ is represented by a vector of weighted term frequency as:

$$d_i = (wt_{i1}, wt_{i2}, \ldots, wt_{in}) \tag{3}$$

where $n$ means the total number of different terms present in a dataset.

Thus, the vector space model can represent a matrix named as weighted-document matrix where terms weights $wt_{ij}$ are in the columns and the documents $d_i$ are in the rows. This weighted matrix gives an input to the clustering algorithms to make an assignment of each document into appropriate clusters. Thus, to determine the similarity between the documents, the proposed method employs the cosine similarity measure.

Given two document vectors $d_i$ and $d_j$, their cosine similarity is computed as [10]

$$\cos(d_i, d_j) = \frac{d_i \cdot d_j}{\|d_i\| \|d_j\|} \tag{4}$$

where $\cdot$ denotes the dot product of the vectors and $\| \ \|$ denotes the length of vector.

## 2.2 The Proposed Algorithm

In this article, genetic algorithm is used for the purpose of determining appropriate initial cluster centers for $K$-means that enhance their performance in text clustering. In GA [11], the initial population of chromosomes has set up at random. At first, we need to define the chromosomes containing $k$ cluster centroids of the initial clusters that are initialized to k different randomly chosen documents from the text corpus. Thus, a chromosome chr as a vector of $k$ cluster centroids is defined as:

$$chr = (cen_1, cen_2, \ldots, cen_k) \tag{5}$$

The objective of fitness function $f_{max}$ is that documents in same group having higher similarity than documents belonging to different group, i.e. maximizing the intra-cluster similarity $f_1$ and minimizing the inter-cluster similarity $f_2$, which are computed as:

$$f_1 = \sum_{j=1}^{k} \left\{ \frac{\sum_{i \in I_j} \frac{x_i \cdot z_j}{\|x_i\| \|z_j\|}}{n_j} \right\} \tag{6}$$

$$f_2 = \sum_{i=1}^{k-1} \sum_{j=i+1}^{k} \left\{ \frac{z_i \cdot z_j}{\|z_i\| \|z_j\|} \right\} \tag{7}$$

Thus, the fitness of chromosomes is measured as:

$$f_{max} = \max(f_1/f_2) \tag{8}$$

where $I_j$ denotes the collection of documents vector belonging to cluster $j$, $z_j$ and $z_i$ denotes the centroids vector of cluster $j$ and cluster $i$ respectively, $n_j$ defines the number of documents in cluster $j$ and $k$ defines the number of clusters.

In genetic algorithm [12], fitness value associated with each chromosome helps in selection process. The proposed method employs a roulette wheel as the selection operator which selects the fittest chromosomes for crossover. This method adopts the uniform crossover which creates the child chromosomes for the next generation. Mutation is done by taking the concept of Gaussian distribution [13]. The process of GA is ended when a predetermined number of iterations reached (Fig. 1). The genetic algorithm can be summarized as:

1. First generate an initial random population of chromosomes, each containing $k$ clusters centroids.
2. Compute the fitness $f(.)$ of each chromosome in the population.
3. Select parent chromosomes from population according to their fitness.
4. Apply crossover operator over selected pair of parents.
5. Perform mutation and get the new population.
6. Determine the fitness of each chromosome in new population.
7. Repeat Steps 3–6 until termination condition is reached.
8. Return with $k$ clusters centroids vectors.

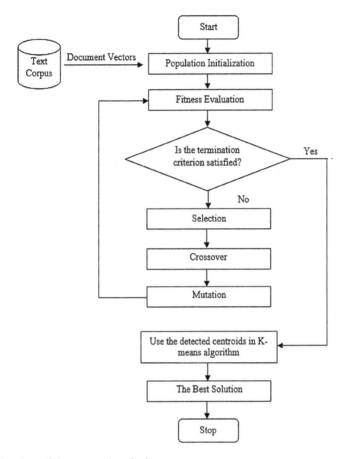

**Fig. 1** Flowchart of the proposed method

## 3  Performance Evaluation

Experiments have been performed on two different Well-known benchmark text datasets, namely, Classic [14] and 20 Newsgroups [15]. Classic dataset containing four different document collection such as CACM, CISI, MED, and CRAN where CACM have 3204 documents, CISI have 1460 documents, MED have 1033 documents and CRAN have 1398 documents. To compare the proposed work with previous work, 2000 documents have considered.

In 20 Newsgroups dataset, data is systematized into 20 different Newsgroups where 1000 Usenet articles were taken from each of 20 Newsgroups. Therefore, dataset consists of 20,000 documents taken from 20 Newsgroups.

For text clustering, we set the GA parameters such as crossover probability 0.6 and mutation probability 0.01. The process of GA ended after a 100 number of iterations.

To evaluate the performance of clustering algorithms, purity and $F$-measure [16, 17] are used, having maximum value 1, which defines the maximum accuracy.

**Purity**

Purity is one of external quality measure to evaluate the quality of clustering. The purity is measured as:

$$\text{Purity} = \frac{1}{N} \sum_k \max_j |w_k \cap c_j| \qquad (9)$$

where $w_k$ represent a cluster $k$, $c_j$ defines the classification having maximum count for cluster $w_k$ and $N$ means the number of documents.

**$F$-measure**

Second external clustering quality measure is $F$-measure. The value of $F$-measure [5] is computed using precision and recall values as below:

$$F\text{-measure}(i,j) = \frac{(2 * \text{Precision}(i,j) * \text{Recall}(i,j))}{(\text{Precision}(i,j) + \text{Recall}(i,j))} \qquad (10)$$

where precision is defined as the ratio of number of members of class $i$ in cluster $j$ and the number of members of cluster $j$ and recall is the ratio of number of members of class $i$ in cluster $j$ and the number of members of class $i$.

Figures 2 and 3 show the purity and $F$-measure results of two clustering algorithms. As it can be seen, the proposed method performs finer and converges to more accurate clusters than the $K$-means method.

**Fig. 2** Purity results of two different algorithms when run on classic and 20 newsgroups datasets

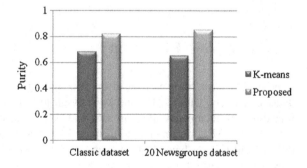

**Fig. 3** *F*-measure results of two different algorithms when run on classic and 20 newsgroups datasets

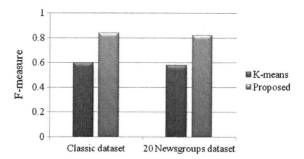

# 4 Conclusion and Future Work

This study presents a document clustering algorithm, where GA is executed at the initial stage to discover the initial cluster centers. The *K*-means algorithm is subsequently applied for refining and creating the final clusters solutions. The clustering results on text datasets show that the proposed method has outperformed the *K*-means method in terms of accuracy. The proposed technique can be extended to make an algorithm that is fully automatic and non-parametric.

# References

1. Konchady, M.: Text mining application programming. Programming Series Charles River Media (2006)
2. MacQueen, J.B.: Some methods for classification and analysis of multivariate observations. In: Proceedings of 5th Berkeley Symposium on Mathematical Statistics and Probability, Berkeley, University of California Press, vol. 1, pp. 281–297 (1967)
3. Han, J., Kamber, M.: Data mining: concepts and techniques, 2nd edn. In: Gray, J. (ed.) The Morgan Kaufmann Series in Data Management Systems. Morgan Kaufmann Publishers (2006)
4. Jain, A.K., Dubes, R.C.: Algorithms for Clustering Data. Prentice-Hall Advanced Reference Series. Prentice-Hall, New Jersey (1988)
5. Steinbach, M., Karypis, G., Kumar, V.: A comparison of document clustering techniques. Technical report, Department of Computer Science and Engineering, University of Minnesota (2000)
6. Sihag, V.K., Kumar, S.: Graph based text document clustering by detecting initial centroids for K-means. Int. J. Comput. Appl. **62**(19) (2013)
7. Premalatha, K., Natarajan, A.M.: Genetic algorithm for documents clustering with simultaneous and ranked mutation. Mod. Appl. Sci. **3**(2), 35–42 (2009)
8. Selim, S.Z., Ismail, M.A.: K-means type algorithms: a generalized convergence theorem and characterization of local optimality. IEEE Trans. Pattern Anal. Mach. Intell. **6**(1), 81–87 (1984)
9. Salton, G., Wong, A., Yang, C.S.: A vector space model for automatic indexing. Commun. ACM **18**(11), 613–620 (1975)

10. Muflikhah, L., Baharudin, B.: Document clustering using concept space and cosine similarity measurement. In: International Conference on Computer Technology and Development, IEEE, vol. 1, pp. 58–62 (2009)
11. Holland, J.: Adaptation in Natural and Artificial Systems. University of Michigan Press (1975)
12. Goldberg, D.E.: Genetic Algorithms in Search, Optimization and Machine Learning. Addison Wesley Publishing Company (1989)
13. Yao, X., Liu, Y., Lin, G.: Evolutionary programming made faster. IEEE Trans. Evol. Comput. 3(2), 82–102 (1999)
14. Classic Dataset. http://www.dataminingresearch.com/index.php/2010/09/classic3-classic4-datasets
15. Newsgroups Dataset. http://archive.ics.uci.edu/ml/datasets/Twenty+Newsgroups
16. Abraham, A., Das, S., Konar, A.: Document clustering using differential evolution. In: IEEE Congress on Evolutionary Computation CEC, pp. 1784–1791 (2006)
17. Aliguliyev, R.M.: Clustering of document collection—a weighting approach. Expert Syst. Appl. 36(4), 7904–7916 (2009)

# Modeling Stochastic Dynamic Facility Layout Using Hybrid Fireworks Algorithm and Chaotic Simulated Annealing: A Case of Indian Garment Industry

Akash Tayal and Surya Prakash Singh

**Abstract** Layout of manufacturing industries is affected by uncertainties such as seasonality, product variation, machine breakdown and other sociopolitical challenges. These uncertainties highly impact the productivity, inventory cost, delivery time, performance quality, and customer satisfaction. To efficiently design the layout considering these uncertainties, it is modeled as Stochastic Dynamic Facility Layout Problem (SDFLP). In this paper, a case of garment manufacturing industry is discussed and solved as a SDFLP. A hybrid Firefly Algorithm (FA) and Chaotic Simulated Annealing (CSA) are used to optimally design layout for the garment manufacturing unit. It is found that the proposed layout is better than the existing layouts.

**Keywords** Chaotic simulated annealing (CSA) · Firefly algorithm (FA) · Facility layout · Meta-heuristic · Stochastic dynamic facility layout

## 1 Introduction

Layout is a design problem where the machines are arranged in the most optimal locations to optimize the material handling cost (MHC). The MHC needs to be optimized not only in single period but also for all periods. The optimization of layout is also influenced by the fluctuations in demand. Therefore, the layout design needs also to take into account the product demand fluctuations. Such layouts are known as SDFLP. Tompkins et al. [1] showed that total MHC is considered to

A. Tayal (✉)
Indira Gandhi Delhi Technical University for Women, Delhi, India
e-mail: akashtayal@yahoo.com

S.P. Singh
Department of Management Studies, Indian Institute of Technology Delhi, New Delhi, India
e-mail: surya.singh@gmail.com

© Springer Nature Singapore Pte Ltd. 2018       31
R.K. Choudhary et al. (eds.), *Advanced Computing
and Communication Technologies*, Advances in Intelligent Systems
and Computing 562, https://doi.org/10.1007/978-981-10-4603-2_4

evaluate the layout efficiency and is 20–50% of total manufacturing cost. Hence, designing an efficient layout can reduce the product cost by 10–30%. The paper presents an application of SDFLP to an Indian Garment shirt manufacturing industry where the products are highly volatile and seasonal. Due to this, the layout needs to be designed optimally by considering product demand variations and rearrangement cost ($RA_c$) of the machines. Therefore, the optimal design of layout becomes a vital issue for such industry where products are highly seasonal and volatile. More details about layout types and its application can be referred from Rosenblatt and Kropp [2], Braglia et al. [3], Kulturel-Konak et al. [4], Singh and Sharma [5, 6], Moslemipour and Lee [7], Matai et al. [8, 9], Tayal and Singh [10–14]. The layout of the garment industry is solved using hybrid FA/CSA algorithm proposed by Tayal and Singh [13].

The paper is structured as follows. Section 2 briefly discusses the SDFLP formulation while Sect. 3 explains hybrid FA/CSA. Section 4 gives the description of the case study on Shirt manufacturing facility. Section 5 gives result and discusses the suggestions for improving the current facility setup. Section 6 concludes the case study.

## 2 SDFLP Model

The SDFLP model is presented in this section where the products demand flows between facilities are uncertain. This uncertainty is represented as probability distribution function (PDF) having certain known value for mean and variance. The mathematical formulation of SDFLP is given below with the notations shown in Table 1 referred from Tayal and Singh [14].

The mathematical formulation used for modeling the shirt manufacturing facility as SDFLP is given in Eq. (1) subject to conditions given in Eqs. (2)–(5). More discussions related to SDFLP can be referred from Kulturel-Konak et al. [4], Moslemipour and Lee [7], Tayal and Singh [10, 14].

$$
\text{Minimize TMHC} = \left\{
\begin{array}{l}
\left[ \sum_{t=1}^{T}\sum_{i=1}^{N}\sum_{j=1}^{N}\sum_{k=1}^{N} \frac{E(D_{tk})}{B_k} C_{tk} \sum_{l=1}^{N}\sum_{q=1}^{N} d_{lq}x_{til}x_{tjq} \right. \\
\left. + Z_p \sqrt{ \sum_{t=1}^{T}\sum_{i=1}^{N}\sum_{j=1}^{N}\sum_{k=1}^{K} \frac{\text{Var}(D_{tk})}{B_k^2} C_{tk}^2 \left( \sum_{l=1}^{N}\sum_{q=1}^{N} d_{lq}x_{til}x_{tjq} \right)^2 } \right] \\
+ \left[ \sum_{t=2}^{T}\sum_{i=1}^{N}\sum_{l=1}^{N}\sum_{q=1}^{N} a_{tilq}x_{(t-1)il}x_{tiq} \right]
\end{array}
\right\}
$$

$$(1)$$

**Table 1** Notations

| Notations | Description |
|---|---|
| $i, j$ | Index for facilities |
| $l, q$ | Index of locations |
| $f_{ij}$ | Material flow between facilities |
| $d_{lq}$ | Distance between locations |
| $N$ | Facility size |
| $C$ | Total MHC of $\pi$ layout |
| $E$ | Expected value of $\pi$ layout |
| Var | Layout variance |
| Pr | Layout probability |
| $Z_p$ | Standard $Z$ considering percentile $p$ |
| $K$ | Part indices |
| $M_{ki}$ | Operation indicator for process on part $k$ using facility $i$ |
| $D_{kt}$ | Demand of part $k$ at $t$ period |
| $B_k$ | Batch size of part $k$ |
| $C_{tk}$ | Movement cost of part $k$ at period $t$ |
| $a_{tilq}$ | Shifting cost |

Subject to:

$$\sum_{i=1}^{N} x_{til} = 1; \quad \forall t, l \tag{2}$$

$$\sum_{l=1}^{N} x_{til} = 1; \quad \forall t, i \tag{3}$$

$$x_{til} = \begin{cases} 1, & \text{if facility } i \text{ located at location } l \text{ in period } t \\ 0, & \text{Else} \end{cases} \tag{4}$$

$$\left| M_{ki} - M_{kj} \right| = 1 \tag{5}$$

Objective function value (OFV) is represented by Eq. (1) where the first term represents MHC while second term represents $RA_c$.

## 3  Hybrid FA/CSA

Meta-heuristics are high level procedure or set algorithm designed to provide good solution for combinatorial optimization problems [10, 15, 16]. In this direction, one of the meta-heuristics proposed by [13] is briefly explained here. Figure 1 presents Hybrid FA/CSA algorithm.

**Table 2** Machine sequence for the product as per the process given in Figs. 2 and 3

| | Product type | Batch size | Machine sequence |
|---|---|---|---|
| Product # 1 | T-Shirt size S (half) | 35 | 1 → 2 → 4 → 5 → 6 → 7 → 10 → 3 → 8 → 12 → 9 → 11 → 13 → 17 → 18 → 19 → 20 |
| Product # 2 | T-Shirt size M (half) | 50 | 1 → 2 → 4 → 5 → 6 → 7 → 10 → 3 → 8 → 12 → 9 → 11 → 13 → 17 → 18 → 19 → 20 |
| Product # 3 | T-Shirt size L (half) | 40 | 1 → 2 → 4 → 5 → 6 → 7 → 10 → 3 → 8 → 12 → 9 → 11 → 13 → 17 → 18 → 19 → 20 |
| Product # 4 | T-Shirt size XL (half) | 25 | 1 → 2 → 4 → 5 → 6 → 7 → 10 → 3 → 8 → 12 → 9 → 11 → 13 → 17 → 18 → 19 → 20 |
| Product # 5 | Shirt size S | 30 | 4 → 1 → 5 → 6 → 2 → 12 → 3 → 7 → 8 → 10 → 9 → 14 → 15 → 13 → 17 → 18 → 19 → 20 |
| Product # 6 | Shirt size M | 60 | 4 → 1 → 5 → 6 → 2 → 12 → 3 → 7 → 8 → 10 → 9 → 14 → 15 → 13 → 17 → 18 → 19 → 20 |
| Product # 7 | Shirt size L | 40 | 4 → 1 → 5 → 6 → 2 → 12 → 3 → 7 → 8 → 10 → 9 → 14 → 15 → 13 → 17 → 18 → 19 → 20 |
| Product # 8 | Shirt size XL | 20 | 4 → 1 → 5 → 6 → 2 → 12 → 3 → 7 → 8 → 10 → 9 → 14 → 15 → 13 → 17 → 18 → 19 → 20 |
| Product # 9 | Shirt embroidery size S | 30 | 16 → 4 → 1 → 5 → 6 → 2 → 12 → 3 → 7 → 8 → 10 → 9 → 14 → 15 → 13 → 17 → 18 → 19 → 20 |
| Product # 10 | Shirt embroidery size M | 50 | 16 → 4 → 1 → 5 → 6 → 2 → 12 → 3 → 7 → 8 → 10 → 9 → 14 → 15 → 13 → 17 → 18 → 19 → 20 |
| Product # 11 | Shirt embroidery size L | 30 | 16 → 4 → 1 → 5 → 6 → 2 → 12 → 3 → 7 → 8 → 10 → 9 → 14 → 15 → 13 → 17 → 18 → 19 → 20 |
| Product # 12 | Shirt embroidery size XL | 25 | 16 → 4 → 1 → 5 → 6 → 2 → 12 → 3 → 7 → 8 → 10 → 9 → 14 → 15 → 13 → 17 → 18 → 19 → 20 |

Stage 1: **Firefly algorithm to find the initial solution**

FA is a population-based algorithm designed on the flashing characteristic of the firefly, which is a function of intensity, $I_p$. FA is applied to get the global solution in wide-search space of SDFLP. Here, each firefly represents a SDFLP solution (layout and its TMHC). Firefly 2, as shown below, exemplifies layout($s_2$), whose TMHC is '$f(s_2)$' (evaluated using Eq. 1) and intensity $I_p = \frac{1}{f(s_2)}$.

$$\text{layout}(s_2) = \begin{cases} N = 1, 2, 3, 4, 5, 6, 7, 8, 9, 10, 11, 12 \\ t = 1[11, 12, 9, 10, 2, 6, 5, 8, 3, 7, 4, 1] \\ t = 2[6, 1, 2, 12, 7, 4, 3, 11, 8, 10, 9, 5] \\ t = 3[3, 7, 9, 8, 4, 2, 12, 5, 1, 10, 11, 6] \\ t = 4[9, 8, 10, 5, 7, 1, 12, 4, 3, 11, 2, 6] \\ t = 5[9, 12, 4, 2, 10, 5, 8, 1, 6, 3, 7, 11] \end{cases}$$

**Stage 1:**
Initialization for FA
    Initialize the population of the fireflies $s_p$ $(p=1, 2, 3...numFF)$

    Each firefly position represents a solution in the search space, $s = \left[s_1, s_2, ..., s_p, ..., s_{numFF}\right]^T$
    Find the firefly light intensity, $I_p$ at $s_p$
    Define the light absorption coefficient, $\gamma$
Firefly Algorithm
    **while** (z < *MaxGeneration*)
        **for** p = 1 : *number of fireflies*
            **for** m = 1 : *number of fireflies*
                **if** $(I_m > I_p)$
                Move firefly 'p' towards firefly 'm'
                Evaluate new solution of firefly 'p' using Equation (1), update $I_p$
                **end if**
                Vary attractiveness with distance r via exp [−γr]
            **end for** m
        **end for** p
        Rank the fireflies and find the current best, z=z+1
    **end while**
Output: The best solution from Firefly Algorithm, s0
**Stage 2:**
Initialization for CSA
    Set initial solution, s0 and assign s=s0
    Generate the chaotic variables using sine map
    Initialize the temperature, $T^0$
**while** $(T^0 <$ Minimum Temperature)
    Generate new neighbourhood solution, s' using Equation (1)
    Compute the OFV i.e. f(s) for s and s'
    lp = 0
    **while** (lp < Inner loop criteria)
        **if**        f(s) > f(s'), assign s = s'
        else **if**   P((f(s')-f(s))/KT) < rand, assign s = s'
        lp = lp + 1
    **end while**
    Decrease $T^0$, using exponential cooling
**end while**
Output: The best solution 's' of SDFLP using Hybrid FA/CSA approach, its Objective Function Values

**Fig. 1** Hybrid FA/CSA algorithm. *Source* Tayal and Singh [13]

Stage 2: **CSA to find the final layout**

CSA is used as an improvement meta-heuristic. The good quality feasible solution from Stage 1 is considered as an input to further improve the final layout and its TMHC. CSA with Sine chaotic map gives the best results in solving SDFLP established [14].

# 4  Case Study of Shirt Manufacturing Industry

ABC Private Limited is a manufacturing firm having operation in Delhi, India. The facility is involved in manufacturing of shirts for various vendors both national and international. Three types of shirts are manufactured for various order size (S, M, L, XL), whose process is shown in Figs. 2 and 3,

1. T-shirt with round neck and half sleeves,
2. Formal shirt without embroidery with full sleeves, and
3. Formal shirt with embroidery on yoke with full sleeves

A garment manufacturing process can be broadly divided into two categories, pre-production and production. The pre-production includes designing of product, pattern and sample design, and grading. The production consists of stitching and finishing. This case study considers the production process with the end objective to find an optimal flexible layout.

The data used in the case study is collected from the secondary sources and derived after discussion with the operation manger. It was seen that the product demand is uncertain during the year owing to the seasonal effect, which is divided into 4 quarters, i.e., time period, $T = 4$. The existing layout at the manufacturing unit was fixed for all time period and the machine was placed in rows of equal dimension. After discussion with operation manager and analysis of data, it was concluded that the case of garment manufacturing facility can be modeled as

---

Shoulder joint → Neck joint → Rib Track →Main label join → Neck tape joint → Neck top sin → Sleeve hamming → Sleeve join → Care label join → Side seam → Sleeve tack → Body hamming → Thread cutting and sucking → Washing → Dry → Iron → Folding and Packing

---

**Fig. 2** Production process for Men's T-Shirt

---

Yoke attach → Shoulder attach → Sleeve tack → Sleeve join → Sleeve top attach → Top stitch arm hole and side seam → Collar attach → Collar close and main label → Embroidery* → Cuff attach → Body hamming → Care label join → Button hole → Button stitching → Thread cutting and sucking → Washing → Dry → Iron → Folding and Packing

---

**Fig. 3** Production process for Men's Formal Shirt with/without* Embroidery

FLP. Here, due to market uncertainty the product demand variation changes the material flow. Thus, the FLP of garment industry can be mapped to SDFLP. For the case study, mapped as a SDFLP, following assumption are considered,

1. Product demand to be Gaussian
2. Facility layout of equal size
3. 'U'-shaped layout is considered
4. Cut material with approved design and marking is available for production of shirts

There are a total of 12 products ($K = 12$) with movement cost 10 for each of them and 20 machines ($N = 20$), Table 3. The product demand is given in Table 4 and machine sequence for product is given in Table 2.

## 5 Case Study Result and Discussion

The existing current layout of garment industry is shown in Fig. 4. To propose new layout for the garment industry considering SDFLP, hybrid FA/CSA proposed by Tayal and Singh [14] is used. The coding is done using Java 8 and it was run on a system with configuration of Intel core 2.53 GHz CPU and 8 GB RAM.

**Table 3** Rearrangement cost of machines of garment industry

| # | Machine type | Rearrangement cost |
|---|---|---|
| Machine # 1 | Overlock machine | 200 |
| Machine # 2 | Overlock machine | 200 |
| Machine # 3 | Overlock machine | 200 |
| Machine # 4 | Plain machine | 100 |
| Machine # 5 | Plain machine | 100 |
| Machine # 6 | Plain machine | 100 |
| Machine # 7 | Plain machine | 100 |
| Machine # 8 | Plain machine | 100 |
| Machine # 9 | Plain machine | 100 |
| Machine # 10 | Flat lock machine | 500 |
| Machine # 11 | Flat lock machine | 500 |
| Machine # 12 | Feed of arm machine | 500 |
| Machine # 13 | Thread cutting and sucking machine | 800 |
| Machine # 14 | Button hole making machine | 500 |
| Machine # 15 | Button stitching machine | 100 |
| Machine # 16 | Zig Zag embroidery machine | 500 |
| Machine # 17 | Washing machine | 1000 |
| Machine # 18 | Tumble dry machine | 1000 |
| Machine # 19 | Ironing machine | 200 |
| Machine # 20 | Packaging machine | 500 |

**Table 4** Product demand

| | Time Period 1 | | Time Period 2 | | Time Period 3 | | Time Period 4 | |
|---|---|---|---|---|---|---|---|---|
| | January–March 2016 | | April–June 2015 | | July–September 2015 | | October–December 2015 | |
| | Mean | Variance | Mean | Variance | Mean | Variance | Mean | Variance |
| Product # 1 | 5216 | 2062 | 3436 | 2132 | 1656 | 2202 | 1784 | 1698 |
| Product # 2 | 10,516 | 2314 | 6366 | 1732 | 2216 | 1151 | 2098 | 1837 |
| Product # 3 | 9365 | 836 | 5962 | 1294 | 2558 | 1752 | 1325 | 1726 |
| Product # 4 | 6310 | 989 | 3986 | 1456 | 1663 | 1922 | 1067 | 920 |
| Product # 5 | 3078 | 2259 | 4089 | 1948 | 5100 | 1636 | 3864 | 2070 |
| Product # 6 | 5465 | 1334 | 8731 | 1143 | 11,998 | 952 | 11,695 | 1338 |
| Product # 7 | 4202 | 1497 | 7406 | 1706 | 10,610 | 1916 | 10,034 | 1822 |
| Product # 8 | 2480 | 1258 | 4325 | 1183 | 6169 | 1107 | 5848 | 2305 |
| Product # 9 | 1927 | 942 | 3368 | 1470 | 4808 | 1999 | 3400 | 1754 |
| Product # 10 | 2480 | 1258 | 4932 | 1264 | 7384 | 1270 | 4916 | 1171 |
| Product # 11 | 2141 | 1837 | 3721 | 1847 | 5300 | 1858 | 4246 | 1506 |
| Product # 12 | 1573 | 2106 | 3045 | 2086 | 4517 | 2067 | 2632 | 1049 |

$$\text{Current layout} = \begin{cases} t = 1[16,4,1,5,6,2,12,3,7,8,10,9,11,14,15,13,17,18,19,20] \\ t = 2[16,4,1,5,6,2,12,3,7,8,10,9,11,14,15,13,17,18,19,20] \\ t = 3[16,4,1,5,6,2,12,3,7,8,10,9,11,14,15,13,17,18,19,20] \\ t = 4[16,4,1,5,6,2,12,3,7,8,10,9,11,14,15,13,17,18,19,20] \end{cases}$$

MHC= 33256514; Rearrangement (RA$_c$)=0; TMHC (MHC+RA$_c$)= 33256514

**Fig. 4** Current layout of the shirt manufacturing garment industry

$$layout(s) = \begin{cases} t = 1[1,4,2,5,6,9,12,14,11,15,16,13,8,3,10,7,17,18,20,19] \\ t = 2[4,1,5,13,6,2,12,3,7,11,16,10,8,9,14,15,17,18,20,19] \\ t = 3[4,1,5,6,2,12,3,7,8,16,10,9,14,11,15,13,17,18,19,20] \\ t = 4[1,5,6,2,12,3,7,8,10,9,14,15,13,16,20,19,17,18,11,4] \end{cases}$$

MHC= 18632762; Rearrangement (RA$_c$)= 16425; TMHC (MHC+RA$_c$)= 18649187

**Fig. 5** Proposed layout from hybrid FA/CSA of shirt manufacturing garment industry

The proposed layout is shown in Fig. 5. The numerical results capture the TMHC solution quality and showed a reduction of 43% in the THMC. Also, the following suggestion were proposed to the manufacturer,

1. The setup should be U-shaped layout and flexible for the 4 quarters
2. Product demand should be considered as Gaussian distribution
3. Rearrangement cost should be included to so as to capture the variance in the demand over the year
4. Cost of movement of the unfinished product from one machine to another need to be included

# 6 Conclusion

In the case study, the existing layout was set up considering average product demand and experience of the operation manager. On analysis of the case, it was concluded that the problem can be mapped to a SDFLP. The paper proposes a flexible layout for the garment manufacturing industry with multi product ($K = 12$) where the product demand is seasonal, i.e., uncertain and multi-period (4 seasons/quarterly) and the number of machines is 20. The model optimizes the TMHC and shows 43% reduction in cost when compared to the cost of the fixed layout used in the existing facility. This supports what Tompkins et al. postulated [1], that if an efficient layout is designed for a manufacturing setup the manufacturing cost can be considerably reduced. This being a case of small manufacturing enterprise, the results can be fine-tuned by the operation manager as per their requirements. To make the evaluation comprehensive, sustainable and realistic, the case can consider the following:

1. Pre-production analysis,
2. Inventory management, and,
3. Qualitative criteria such as, safety, maintenance, adjacency, waste disposal, etc., which can make the formulation to be multi-objective.

# References

1. Tompkins, J.A., White, J.A., Tanchoco, J.M.A.: Facilities Planning. Willey, New York (2003)
2. Rosenblatt, M.J., Kropp, D.H.: The single period stochastic plan layout problem. IIE Trans. **24**(2), 169–176 (1992)
3. Braglia, M., Zanoni, S., Zavanella, L.: Layout design in dynamic environments: strategies and quantitative indices. Inte. J. Prod. Res. **41**(5), 995–1016 (2003)
4. Kulturel-Konak, S., Smith, A.E., Norman, B.A.: Layout optimization considering production uncertainty and routing flexibility. Int. J. Prod. Res. **42**(21), 4475–4493 (2004)
5. Singh, S.P., Sharma, R.R.K.: A review of different approaches to the facility layout problem. Int. J. Adv. Manuf. Technol. **30**(5–6), 425–433 (2006)
6. Singh, S.P., Sharma, R.R.K.: Two level simulated annealing based approach to solve facility layout problem. Int. J. Prod. Res. **46**(13), 3563–3582 (2008)
7. Moslemipour, G., Lee, T.S.: Intelligent design of a dynamic machine layout in uncertain environment of flexible manufacturing systems. J. Intell. Manuf. **23**(5), 1849–1860 (2011)
8. Matai, R., Singh, S.P., Mittal, M.L.: Modified simulated annealing based approach for multi objective facility layout problem. Int. J. Prod. Res. **51**(14), 4273–4288 (2013)
9. Matai, R., Singh, S.P., Mittal, M.L.: A new heuristic for solving facility layout problem. Int. J. Adv. Oper. Manag. **5**(2), 137–158 (2013)
10. Tayal, A., Singh, S.P.: Chaotic simulated annealing for solving stochastic dynamic facility layout problem. Int. J. Manag. Stud. **14**(2), 67–74 (2014)
11. Tayal, A., Singh, S.P.: Integrated SA-DEA-TOPSIS based solution approach for multi objective stochastic dynamic facility layout problem. Int. J. Bus. Syst. Res. (Accepted) (2016)

12. Tayal, A., Singh, S.P.: Analysis of simulated annealing cooling schemas for design of optimal flexible layout under uncertain dynamic product demand. Int. J. Oper. Res. (in press) (2016)
13. Tayal, A., Singh, S.P.: Designing flexible stochastic dynamic Layout: An integrated firefly and chaotic simulated annealing based approach. Int. J. Flex. Syst. Manag. (Accepted and available online) (2016)
14. Tayal, A., Singh, S.P.: Analysing the effect of chaos functions in solving stochastic dynamic facility layout problem using CSA. Advances in Intelligent Systems and Computing, 99–108. Springer, Singapore (2016)
15. Mingjun, J., Huanwen, T.: Application of chaos in simulated annealing. Chaos, Solitons Fractals **21**, 933–941 (2004)
16. Yang, X.S.: Firefly algorithms for multimodal optimization. Stochastic Algorithms: Foundations and Applications, pp. 169–178. Springer, Berlin (2009)

# Transition-Aware Human Activity Recognition Using eXtreme Gradient Boosted Decision Trees

Kunal Gusain, Aditya Gupta and Bhavya Popli

**Abstract** Gradient Boosted Machines (GBM) have long been used for regression and classification purposes, and their basic premise of boosting weak learners has been adapted and improvised time and time again in search for an optimal modeling approach. One recent variant has taken the world of learning-related problems by storm, by giving the most accurate solutions to most of them, it is called eXtreme GBM. The paper evaluates eXtreme GBM in the context of human activity recognition.

**Keywords** Machine learning · Gradient boosted machines · XGBoost · Human activity recognition · Transition-Aware

## 1 Introduction

In many real-world scenarios, for accessing of big data from storages and online learning problems, we often do not have the luxury of training models on the entire dataset. The data arrives in batches or small sets and machines have to be trained on these reduced inputs. This however is not necessarily a bad thing, since machines when capable of effective classification even from a small amount of data at a time, can help reduce the problem of large memory requirement and prevent models from being computationally intractable. Support vector machines (SVM) are proven efficacious classifiers in many fields [1, 2], and their usage in incremental learning has shown great results [3, 4]. Not only do they learn from the previous classification, they even

K. Gusain (✉) · A. Gupta · B. Popli
Computer Science Department, Bharati Vidyapeeth's College of Engineering,
Guru Gobind Singh Indraprastha University, Delhi, India
e-mail: kunalgusain1995@gmail.com

A. Gupta
e-mail: adityag95@gmail.com

B. Popli
e-mail: bhavyapopli096@gmail.com

© Springer Nature Singapore Pte Ltd. 2018
R.K. Choudhary et al. (eds.), *Advanced Computing
and Communication Technologies*, Advances in Intelligent Systems
and Computing 562, https://doi.org/10.1007/978-981-10-4603-2_5

possess the ability to incorporate new classes in new batches. Ensemble of SVMs trains the machines in an incremental fashion, where the previously trained machine is used for classifying the next sample, and this twice trained machine is used on the next dataset. Multiple realizations of incremental SVMs are possible and proposed methodology makes use of misclassified data and an array of SVMs to further enhance the traditional approaches.

## 2    eXtreme Gradient Boosted Decision Trees

Models for efficient learning have always been needed, and Gradient Boosted Machine (GBM) is one of the best that have come up over time. A powerful algorithm, it finds extensive usage in real-world scenarios [3]. GBMs are essentially an extension of the concept of boosting in machine learning, where we combine up the results of different prediction models (which are mostly Decision Trees) or classifiers by taking their weighted averages, we then iteratively try to learn from the weak prediction models at each step and correct the errors of the previous iteration in the next one. The algorithm can be thought of as a problem of optimizing a cost function (which are usually RMSEs or MSEs) which consists of the losses that occur while fitting the most appropriate model to the data which need to be minimized. This is being done by iteratively choosing a function that points in the direction of the negative gradient. This implies that after the calculation of error function, we take up its negative derivative at each step, fit up a model to the function obtained, and finally add it to the previous model after assigning it the proper weight. This helps in learning from the errors (or losses) which had occurred during the prediction in the previous iteration; this is the basic idea behind the GBM process.

The recently proposed, eXtreme Gradient Boosted Decision Trees are special type of improvised version of decision trees used for efficiently analyzing a given dataset as compared to the other decision tree algorithms. More specifically, these are used in supervised machine learning problems in which we try to predict the outcome of a given test set on the basis of the data which is already available with us. The eXtreme Gradient Boosted Trees have been implemented in an open source software library called XGBoost.[1] They are an extension to the decision trees that use the Gradient Boosting Algorithm and produce customizable models with less overfitting [5, 6].

For understanding the XGBoost Decision Trees, the best way forward would be the official documentation,[2] written by the creators of the library. Building upon the work done by them [5–7], we explain the methodology by first defining optimization function, depicted in (1).

---

[1]https://github.com/dmlc/xgboost.

[2]https://xgboost.readthedocs.io.

$$obj(\Theta) = L(\theta) + \Omega(\Theta) \tag{1}$$

The objective function consists of two terms—The first one is representative of different types of losses that may occur while fitting the model on the dataset, they could be Root Mean Squared Loss or Logistic Losses, and so on. The second term is called the regularization term, which GBMs do not use, this term helps in avoiding overfitting of the model on the training data. For further simplification, we take Root Mean Square Error (RMSE) as our objective function. Mathematically, we can define the tree ensemble model by the mathematical equation given in (2).

$$\hat{y}_i = \sum_{k=1}^{K} f_k(x_i), \quad f_k \in F \tag{2}$$

Here, $K$ refers to the number of trees which we are combining, $F$ refers to the set of all possible trees in space, and $f_k$ is a tree which belongs to $F$. Hence, we can modify the objective function as shown in (3),

$$obj(\theta) = \sum_{i}^{n} l(y_i, \hat{y}_i) + \sum_{k=1}^{K} \Omega(f_k) \tag{3}$$

Our aim is the minimization of the objective function. According to the principle of GBMs [3], we need to find the next weak learner or classifier by adding one new tree at a time, that is, by the usage of an additive strategy. Therefore, if we consider Gradient Boosting process to be an iterative one, then, the tree which is being obtained at the $t$th step in the process would be the sum of the tree that was obtained in $(t-1)$th step, and of some function which is the new weak learner of the dataset. Therefore, if we make this change in the objective function, then, it looks like (4),

$$obj^{(t)} = \sum_{i=1}^{n} l\left(y_i, \hat{y}_i^{(t-1)} + f_t(x_i)\right) + \Omega(f_t) + constant \tag{4}$$

In order to obtain a more simplified form of the objective function, we take its Taylor series expansion. After taking the expansion and dropping all the constants, the objective function now becomes something like in (5),

$$obj^{(t)} = \sum_{i}^{n} \left[ g_i f_t(x_i) + \frac{1}{2} h_i f_t^2(x_i) \right] + \Omega(f_t), \tag{5}$$

where

$$g_i = \partial_{\hat{y}^{(t-1)}} l\left(y_i, \hat{y}^{(t-1)}\right) \quad \text{and} \quad h_i = \partial_{\hat{y}^{(t-1)}}^2 l\left(y_i, \hat{y}^{(t-1)}\right)$$

Here, $g_i$ and $h_i$ are called Gradient and Hessian respectively and they refer to the first- and second-order partial derivative of the loss function at the $(t - 1)$th step with respect to $\widehat{y}$ at the $(t - 1)$th step.

Thus the function in (5) is the optimization cost function for the classifier that we obtain at the $t$th step. For simplifying the regularization term, we will have to reconsider the definition of the weak learner which is being iteratively added at every step during the Gradient Boosting process, i.e., $f_t(x_i)$. Here, we redefine it as shown in (6),

$$f_t(x) = w_{q(x)}, w \in R^T, q : R^d \rightarrow \{1, 2, \ldots, T\}, \tag{6}$$

where $T$ refers to the total number of leaf nodes in the tree, $q(x)$ is a function that maps each of the data point in the training set to a specific leaf node of the weak classifier, and $w$ refers to the vector score (the prediction score) on the leaf. Hence, by considering this definition, the regularization complexity of XGBoost Trees has been defined in (7).

$$\Omega(f) = \sqrt{T} + \frac{1}{2}\lambda \sum_{j=1}^{T} w_j^2 \tag{7}$$

Chen et al.'s official model accepts the fact that there could be more ways to compute the complexity, however the proposed approach has worked wonderfully in practice [5]. Regularization gives XGBoost an edge over other machine learning algorithms, which have traditionally failed to effectively control overfitting and complexity. Introducing this regularization term in the simplified objective function, the final equation we get looks something like in (8),

$$\text{obj}^{(t)} = \sum_{j=1}^{T} \left[ G_j w_j + \frac{1}{2}(H_j + \lambda)w_j^2 \right] + \sqrt{T}, \tag{8}$$

where $I_j = \{i | q(x_i) = j\}$

$$G_j = \sum_{i \in I_j} g_i \quad \text{and} \quad H_j = \sum_{i \in I_j} h_i$$

Here, $I_j$ refers to all the data points which are associated with the $j$th leaf. Now, this is the objective function for the $t$th weak classifier considering all the leaf nodes and we need to minimize this function. After solving the quadratic equation we finally get:

$$w_j^* = -\frac{G_j}{H_j + \lambda}$$

$$\text{obj}^* = -\frac{1}{2}\sum_{j=1}^{T}\frac{G_j^2}{H_j + \lambda} + \sqrt{T}$$

This is the final objective function that has to be minimized for each of the weak classifier and this would determine how the splitting [5, 8] of the nodes would take place at each level. EXtreme Gradient Boosted Decision Trees have been developed both in deep consideration with principles of machine learning and system optimization. XGBoost aims in providing a "Scalable, Portable and Accurate Library" for the eXtreme Gradient Boosted Decision Trees, and continues to give better and more accurate results than many of the traditional approaches.

# 3 Transition-Aware Human Activity Recognition

Awareness regarding a user's body movements, and its effective utilization can lead to immense benefits. Some immediate ones could be in tracking movement, assisting the handicapped, automating actions or sequences, security, ubiquitous computing and in broader sense, the entire spectrum of Wide and Pervasive Computing [9] as well as Ambient Intelligence [2]. Although this field has witnessed quite an extensive amount of research, the problems of efficient data collection and its optimal usage have persisted. The defining work to be done and studied recently was done by Reyes-Ortiz et al. [10] which attached a smartphone to 30 subjects, of varying age groups from 19 years of age to 48 years old, and using the phone monitored there movements and postural transitions. Usage of smartphones, over the erstwhile Accelerometers and Gyroscopes, significantly improves the process of wireless transmission of readings, their evaluation and the presence of precise, and a varied amount of sensors makes sure that efficient recording of data can take place. The dataset has been made available for free to the public on the UCI Machine Learning Repository[3] as well as the website of Human Activity Recognition Labs.[4] The smartphone-based recognition of Human Activities and Postural Dataset (HAPT) makes use of six basic activities—standing, sitting, lying, walking, walking upstairs, and walking downstairs.

---

[3]https://archive.ics.uci.edu/ml/datasets/Smartphone-Based+Recognition+of+Human+Activities+and+Postural+Transitions.

[4]Jorge L. Reyes-Ortiz(1,2), Davide Anguita(1), Luca Oneto(1) and Xavier Parra(2)

1. Smartlab, DIBRIS—UniversitÃ degli Studi di Genova, Genoa (16145), Italy.
2. CETpD—Universitat PolitÃ¨cnica de Catalunya. Vilanova i la GeltrÃº (08800), Spain har'@'smartlab.ws, www.smartlab.ws.

**Table 1** Description of the HAPT dataset

| Type of dataset | Multivariate, time series | Number of instances | 10,929 |
|---|---|---|---|
| Attribute characteristics | Real | Number of attributes | 561 |
| Task at hand | Classification | Missing values | None |

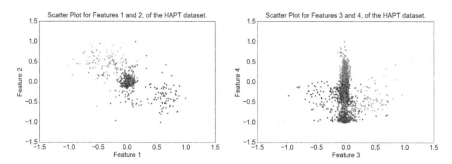

**Fig. 1** Scatter plots of some features from the HAPT dataset

A data with high dimensionality (561 features), it is essentially a time series analysis, and necessitates the use of a strong modeling approach for its classification and recognition. Salient details of the dataset have been given in Table 1. All the three axes of linear acceleration and angular velocity were recorded, for this multivariate dataset, which has been divided into 70 and 30% for the purposes of training and testing, respectively. To better appreciate the features and their groupings, scatter plots of some of them have been visualized using scatter plots in Fig. 1. The paucity of an efficient Transition-Aware recognition System was also one of our motivations to study and find a methodology to effectively classify this. The transitions are between—standing and sitting, standing and lying, and standing and lying. The possibilities for usage of Transition-Aware Recognition Systems (TARS) are boundless, and together with advances in modeling and learning systems, a search for an approach to achieve the best possible implementation, must alas remain endless. Our work on TARS is essentially a search for its most accurate realization, so that it may act as a launch bed for future usage and research.

## 4   Experimental Evaluation

Problems of accuracy with similar datasets have persisted, and thus our aim was to find a more efficient one [11–13].

To obtain TARS, we tested various benchmark algorithms for obtaining the best results along with xgboost algorithms. Xgboost algorithm can be considered as the darling of the recent data science competitions as 90% of the winning solutions of the competitions use xgboost.

1. K-Neighbors Classifier (K-NN) [14].
2. Support Vector Machines (SVM) [15].
3. Linear Discriminant Analysis (LDA) [16].
4. Decision Trees (DT) [17].
5. AdaBoost [18].
6. Random Forests Classifier (RF) [19].
7. Naïve Bayes Classifier (NB) [4].
8. Logistic Regression (LR) [1].
9. Neural Networks (NN).
10. XGBoost.

Figure 2 depicts the performance of all the algorithms on the HAPT dataset. It is clear from Fig. 2 that xgboost algorithm outperforms all the other benchmark algorithms for accurately predicting the activity of the user. Logistic regression comes second in terms of accuracy and adaboost and naïve Bayes performs the worst among all the other algorithms (Table 2).

**Fig. 2** Accuracy comparison of various algorithms

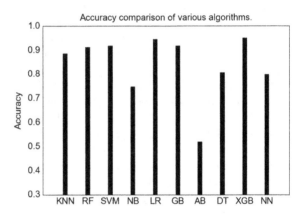

**Table 2** Accuracy table

| Algorithm name | Accuracy |
| --- | --- |
| KNN | 88.5515496521 |
| RF | 91.1448450347 |
| SVM | 91.8089816571 |
| NB | 74.7311827956 |
| LR | 94.5287792536 |
| GB | 91.8089816571 |
| AB | 51.9607843137 |
| DT | 80.7084123972 |
| XGB | 94.9715370019 |
| NN | 80 |

All the algorithms were implemented and executed on the computer with following hardware configuration.

**I7, 8 gbram INCOMPLETE H/W Description**

## 5  Conclusions and Future Scope

An effective approach for classification problems that arrive from online learning is needed. In incremental learning instead of training the machine on the entire dataset, small batches are formed, and then the model works on each dataset, learning from the previous classification without storing the actual data. Incremental SVMs also utilize this approach, and thus we save on both space and time. They, however, suffer from the problem of catastrophic forgetting. The proposed approach implements incremental learning using an ensemble of SVMs. Once the machine is trained on the first batch of data, it is stored in any array of machines. The second time around, this machine is trained on the new batch, and correctly classified data is discarded, but we train a new machine of the misclassified one. Similarly, we keep training machines on subsequent batches and generating SVM for each misclassified batch. Final classification at each step is obtained by taking a weighted sum of all the machines in the array, and these weights are assigned sing PSO. Simulation is run on the Human Activity and Postural Transition dataset to obtain a Transition-Aware Classification. Proposed approach shows a higher accuracy than the traditional one and also runs in significantly lesser time. The results are promising and attest to the veracity of the model put forward.

Future work could focus on comparing the proposed approach with the multitude of Incremental SVM realizations that have come up over time. Simulations on multiple datasets, of varying variety, need to be run and studied to gain more insight. The storing and usage of classifiers used in the previous machines, for the next batch of data, could be studied and their results contrasted with current approaches. Alternatives for optimization can be implemented and studied. Hybrid or improved variants of SVMs using proposed ensemble method may also prove interesting.

## References

1. Hosmer Jr., D.W., Lemeshow, S.: Applied logistic regression. Wiley (2004)
2. Chen, L., Hoey, J., Nugent, C.D., Cook, D.J., Yu, Z.: Sensor-based activity recognition. IEEE Trans. Syst. Man Cybern. PartC: Appl. Rev. **42**, 790–808 (2012)
3. Friedman, J.H.: Greedy function approximation: a gradient boosting machine. Ann. Stat. 1189–1232 (2001)
4. Murphy, K.P.: Naive bayes classifiers. University of British Columbia (2006)

5. Chen, T., Guestrin, C.: XGBoost: a scalable tree boosting system. arXiv: 1603.02754v3 [cs. LG] 10 Jun 2016
6. Chen, T., Guestrin, C.: XGBoost: reliable large-scale tree boosting system
7. Chen, T., He, T.: Higgs boson discovery with boosted trees. In: JMLR: Workshop and Conference Proceedings, vol. 42, pp. 69–80 2015
8. Chen, T.: Introduction to Boosted Trees. University of Washing Computer Science, University of Washington, 22 Oct 2014
9. Cook, D.J., Das, S.K.: Pervasive computing at scale : Transforming the state of the art. Pervasive Mob. Comput. **8**, 22–35 (2012)
10. Reyes-Ortiz, J.L., Oneto, L., Samà A., Parra, X., Anguita, D.: Transition-aware human activity recognition using smartphones. Neurocomputing (2015)
11. Kwapisz, J.R., Weiss, G.M., Moore, S.A.: Activity recognition using cell phone accelerometers. ACM SigKDD Explorations Newsl. **12**, 2–74 (2011)
12. Kwapisz, J.R., Weiss, G.M., Moore, S.A.: Cell phone-based biometric identification. In: 2010 Fourth IEEE International Conference on Biometrics: Theory Applications and Systems (BTAS) (2010)
13. Ravi, N., Dandekar, N., Mysore, P., Littman, M.L.: Activity recognition from accelerometer data. In: AAAI, vol. 5, p.1541 (2005)
14. Altman, N.S.: An introduction to kernel and nearest-neighbor nonparametric regression. Am. Stat. **46**(3), 175–185 (1992)
15. Hearst, M.A., et al.: Support vector machines. IEEE Intell. Syst. Appl. **13**(4), 18–28 (1998)
16. Scholkopft, B., Mullert, K.-R.: Fisher discriminant analysis with kernels. Neural Netw. Sig. Process. IX **1**(1), 1 (1999)
17. Quinlan, J.R.: Induction of decision trees. Mach. Learn. **1**(1), 81–106 (1986)
18. Zhu, J., et al.: Multi-class adaboost. Stat. Interface **2**(3), 349–360 (2009)
19. Breiman, L.: Random forests. Mach. Learn. **45**(1), 5–32 (2001)

# Software Defect Prediction: A Comparison Between Artificial Neural Network and Support Vector Machine

Ishani Arora and Anju Saha

**Abstract** Software industry has stipulated the need for good quality software projects to be delivered on time and within budget. Software defect prediction (SDP) has led to the application of machine learning algorithms for building defect classification models using software metrics and defect proneness as the independent and dependent variables, respectively. This work performs an empirical comparison of the two classification methods: support vector machine (SVM) and artificial neural network (ANN), both having the predictive capability to handle the complex nonlinear relationships between the software attributes and the software defect. Seven data sets from the PROMISE repository are used and the prediction models' are assessed on the parameters of accuracy, recall, and specificity. The results show that SVM is better than ANN in terms of recall, while the later one performed well along the dimensions of accuracy and specificity. Therefore, it is concluded that it is necessary to determine the evaluation parameters according to the criticality of the project, and then decide upon the classification model to be applied.

**Keywords** Back propagation · Supervised learning · Artificial neural networks · Software quality · Support vector machine

I. Arora (✉)
Northern India Engineering College, FC-26, Shastri Park, Delhi 110053, India
e-mail: ishaniarora06@gmail.com

A. Saha
University School of Information and Communication Technology,
Guru Gobind Singh Indraprastha University, Sector-16C, Dwarka,
Delhi 110078, India
e-mail: anju_kochhar@yahoo.com

© Springer Nature Singapore Pte Ltd. 2018
R.K. Choudhary et al. (eds.), *Advanced Computing
and Communication Technologies*, Advances in Intelligent Systems
and Computing 562, https://doi.org/10.1007/978-981-10-4603-2_6

51

# 1   Introduction

Software defect prediction (SDP) is used to assess the software quality by building the software defect prediction models using the static software design and code metrics. SDP is usually treated as a binary classification problem where in a module is either defect prone or nondefect prone. A module is a single indivisible unit of the source code, which may be a class or a procedure, having a set of object-oriented features such as Chidamber and Kemerer metrics [1] or procedural metrics such as Halstead metrics [2], respectively. A software defect prediction model is built through a training phase using the labeled historical defect data and then, a trained model acts as a classifier for the new unknown data. The classification performance of the SDP model is examined along various parameters such as accuracy, sensitivity, and specificity.

Researchers and practitioners have successfully applied the statistical techniques, such as logistic regression [3], and machine learning techniques, specifically supervised learning methods, such as decision trees [4], Naïve Bayes [4], and support vector machines [5], as a solution to the SDP problem. The dependence of the statistical methods on the characteristics of the data set proved them inadequate over the machine learning algorithms. The ability to learn from the data without being explicitly programmed an important feature of machine learning algorithms.

Support vector machine (SVM), a supervised learning technique, has been efficiently applied to solve a majority of classification and regression problems in the recent years [5, 6]. The operating principle behind SVM is the Structural Risk Minimization (SRM) which works with the aim of minimizing the generalization error. Some important characteristics of SVM are:

- Ability to generalize in multidimensional spaces with small amount of training samples.
- Not affected and hence, robust to the outliers.
- Ability to model the nonlinear relationships between the software metrics and the defect proneness.
- Achieves a global optimum solution.

Although the prediction models were effectively built for the classification task using machine learning algorithms, but there was no way that the systems could become intelligent, in the sense that they could not sense the severity of the defect proneness of the module and hence, prioritize the defects. This characteristic of intelligence was inherent with the concept of Artificial Intelligence (AI) introduced as a separate field in the 1990s [7]. With the recent advancements, the domain of machine learning got merged with AI as its subset and hence, the terms are often used interchangeably. AI techniques such as artificial neural networks (ANNs) and fuzzy logic have proved their applicability in the domain of software engineering, such as cost estimation [8], effort estimation [9] and software defect prediction [10, 11].

An artificial neural network is a biologically inspired set of interconnected parallel operating units called neurons. Each neuron takes an input which is the product of the actual input value and a weight. The summation of these individual product values is passed as an input to a transfer function which produces the output for the respective layer. This output may act as an input for the next neuron or may be produced as the final output of the ANN. The basic architecture of a neural network is characterized by three layers: input layer, output layer, and the hidden layer. The advantages of ANNs are manifold: (1) Ability to learn itself and discover knowledge, (2) Ability to model the complex nonlinear relationships between the software metrics and the defect data, and (3) Expandability and simple structure. ANNs have been applied for software defect prediction problem successfully by various researchers, and have shown favorable results [10].

This empirical study is executed with two main objectives:

- To show the classification performance of SDP models built using the static design and code metrics
- An empirical performance comparison of SVM and ANN, and determine which one is better.
- The tasks, therefore, include:
- Building of software defect prediction models
- Compare the results of SVM and ANNs

The classification models were built using public domain data sets from the PROMISE repository [12]. Public domain data sets always allow for the verifiability and refutability of the obtained results. The prediction performance was evaluated along the parameters of accuracy, sensitivity and specificity.

The paper is organized as follows: Sect. 2 provides a brief overview of the background study of SVM and ANNs in the context of SDP. Section 3 presents an overview of SVM and ANN classifiers. Section 4 describes the characteristics of the data as well as the performance evaluation measures used for the experimental study. Results are presented and discussed in Sect. 5. Conclusions are given in Sect. 6.


## 2 Background Study

The existing SDP literature has shown an immense growth of the application of support vector machines and artificial neural networks since their introduction as the classification models.

Lanubile et al. [13] performed a comparison of six classification techniques: discriminant analysis, principal component analysis (PCA), logistic regression (LR), layered neural networks, holographic networks and logical classification models. The neural network model was developed using back propagation learning using 27 academic projects. Misclassification rate, predictive validity, verification cost and achieved quality were used as performance evaluation measures. However, the required performance benchmark could not be attained by any of the SDP models.

Elish and Elish [5] performed an empirical study by building defect prediction models using SVM on four publically available NASA datasets, namely PC1, CM1, KC1 and KC3, and compared its performance with eight statistical and machine learning methods, i.e. logistic regression (LR), decision trees, K-nearest neighbours (KNN), multilayer perceptrons (MLP), radial basis function (RBF), Naïve Bayes, Bayesian belief networks (BBN) and random forest (RF). The performance of all the models was compared along the parameters computed from confusion matrix and it was concluded that SVM's performance was better or at least equivalent to the other SDP techniques.

Gondra [14] performed a study with an objective to determine the significant software metrics for defect estimation. An ANN model was trained using the historical data and then, sensitivity analysis was performed in order to find out the significant metrics. These metrics were further used to build a separate ANN model for determining the defect proneness of each software module. The classification performance of the ANN model was compared against the Gaussian kernel SVM as the standard. The experiment was conducted using JM1 dataset from the NASA MDP repository and the results showed that SVM outperformed ANN in the binary defect classification problem.

Zheng [15] performed an empirical study using three cost-sensitive boosting algorithms and backpropagation learning algorithms. Out of three, one was based on the threshold and the other two were the weight updating architectures. The study was conducted using four NASA datasets and normalized expected cost of misclassification (NECM) as the performance measure. The results suggested that the threshold based feedforward neural network outperformed the other methods, especially for the object oriented software modules.

Malhotra [16] performed a comparative study of statistical and machine learning algorithms for building the SDP models using the public domain data sets of AR1 and AR6. The techniques included decision trees, artificial neural networks, support vector machines, cascade correlation network, group method of data handling method, and gene expression programming. The AUC values were used to confirm the prediction capability of these methods for SDP. The AUC values for the decision tree method were found to be 0.8 and 0.9 for the AR1 and AR6 defect data set, respectively, which was better than the predictive capability of the other techniques in comparison.

The relationships between the software metrics and the probability of the occurrence of the defect are often seen as complex. SVM and ANN are the two identified machine learning techniques which can handle the complexity and nonlinearity of such relationships and therefore, build effective defect prediction models. In this paper, the defect prediction models are built using SVM and ANN using the static design and code metrics for the binary classification problem. Further, the predictive capability of the models is compared in terms of various parameters computed from the confusion matrix as described in Sect. 4.2.

# 3   Techniques Involved

## 3.1   Support Vector Machines

One of the supervised machine learning algorithms which can handle both classification and regression problems efficiently is the support vector machine (SVM).
An SVM model represents the instances as a set of points in the space which are
separated in a way that the different categories are as wide as possible. The predictions are done for the unlabeled data based on which side they fall on. The
functions which perform this mapping into the space are called as kernel functions.
Linear, polynomial, Gaussian, and sigmoid functions are the most common types of
kernel functions [17–19]. A linear SVM classifier places a single data point in a
p-dimensional space and it finds out if these points can be separated with $(p - 1)$
dimensional hyperplane. The optimal choice of the hyperplane is the one which
separates the data points with the largest margin between the two classes. In other
words, a good choice of the separating hyperplane is made which has the maximum
margin from the nearest training data point and hence, minimizes the generalization
error (Fig. 1).

## 3.2   Artificial Neural Networks

The basic architecture of an artificial neural network is shown in Fig. 2. ANNs are a
system of many interconnected processing units called as neurons. An ANN is

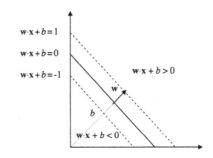

**Fig. 1**  A simple linear
hyperplane classifier

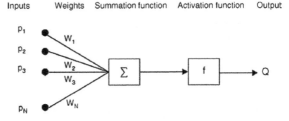

**Fig. 2**  Basic architecture of
an artificial neural network

analogous to the biological model of the brain. An important feature which makes them highly useful is their ability to learn from the data and therefore, adapting themselves in the context in which they are trained. ANNs are, therefore, referred to as "data rich" and "theory poor" models because it does not need any prior knowledge of the problem at hand and sufficient number of instances for training purpose is enough [20]. The most common learning algorithm for ANN is the backpropagation learning algorithm [21]. The backpropagation algorithm uses error as the performance function and optimizes on the minimization of error. The error is usually represented in terms of mean square error (MSE) or sum of squared error (SSE). The default back propagation training algorithm is based on the Levenberg–Marquardt learning function [22].

## 4 Experiments

### 4.1 Data Involved

This comparative study involved seven datasets from the PROMISE repository [12], namely, PC1, PC2, PC3, PC4, PC5, KC2, and KC3. The characteristics of these datasets, such as the number of instances and the percentage of the number of defective modules, are described in Table 1. The software metrics shared by these defect data sets are the Halstead metrics [2], McCabe metrics [1] and the Lines of Code (LOC) metrics as described in Table 2. The parameters defined during the training of the SVM and the ANN techniques to build the software defect prediction models are defined in Tables 3 and 4, respectively.

### 4.2 Performance Measures

Confusion matrix [23], as shown in Fig. 3, is generally used as the standard for determining the performance of classification models. This section describes the various performance evaluation measures which can be computed from the confusion matrix and hence, are used in this work.

Table 1 Characteristics of the data set involved

| Data set | No. of instances | Defect percentage |
|----------|------------------|-------------------|
| PC1 | 565 | 7.61 |
| PC2 | 745 | 2.15 |
| PC3 | 1077 | 12.44 |
| PC4 | 1458 | 12.21 |
| PC5 | 17,186 | 3.00 |
| KC2 | 522 | 20.50 |
| KC3 | 194 | 18.56 |

**Table 2** Common software metrics

| McCabe | Cyclomatic complexity |
| --- | --- |
| | Design complexity |
| | Essential complexity |
| | Normalized cyclomatic complexity |
| Halstead | Num operands |
| | Num operators |
| | Num unique operands |
| | Num unique operators |
| | Error estimate |
| | Length |
| | Level |
| | Program time |
| | Effort |
| | Difficulty |
| | Volume |
| | Content |
| Lines of code (LOC) | Blank |
| | Code and comments |
| | Executable |
| | Comments |
| | Total |

**Table 3** SVM parameters used

| S. No. | Parameter | Value |
| --- | --- | --- |
| 1 | Function | Linear |
| 2 | Solver | SMO |
| 3 | Scale | 1 |
| 4 | Training data: test data | 60:40 |

**Table 4** ANN parameters used

| S. No. | Parameter | Value |
| --- | --- | --- |
| 1 | Training function | trainlm |
| 2 | Performance function | MSE |
| 3 | Maximum number of epochs for training | 1000 |
| 4 | Performance goal | 0 |
| 5 | Maximum validation failures | 6 |
| 6 | Minimum performance gradient | $1-e7$ |
| 7 | Number of neurons in the hidden layer | 10 |

**Fig. 3** Confusion matrix

|  |  | Actual/Target class | |
| --- | --- | --- | --- |
|  |  | No | Yes |
| Predicted/Output class | No | TN | FN |
| | Yes | FP | TP |

where,

TN = True negative = the no. of nondefective modules which are correctly predicted to be nondefective
FN = False negative = the no. of defective modules incorrectly classified as nondefective
FP = False positive = the no. of nondefective modules incorrectly classified as defective
TP = True positive = the no. of defective modules correctly predicted as defective

**Accuracy**

Accuracy is the total number of defective and nondefective modules identified correctly, out of the total number of modules in the software project. In mathematical terms, it is defined as:

$$\text{Accuracy} = \frac{\text{TN} + \text{TP}}{\text{TN} + \text{FP} + \text{FN} + \text{TP}} \tag{1}$$

**Recall**

Recall depicts the ratio of the number of correctly identified faulty modules to the total number of actually faulty modules in the data set. Recall is also called as the sensitivity of the SDP model, and is defined as:

$$\text{Recall} = \frac{\text{TP}}{\text{TP} + \text{FN}} \tag{2}$$

**Specificity**

Specificity is the ratio of the number of correctly identified non-faulty modules to the total number of actually non-faulty modules in the data set.

$$\text{Specificity} = \frac{\text{TN}}{\text{TN} + \text{FP}} \tag{3}$$

## 5 Results

The experiments were executed in the MATLAB 2015 environment. The SVM and multilayer feedforward neural network classification models were constructed using the command line interface. The results obtained on the three parameters described in Sect. 3.2 are shown in Tables 5, 6 and 7.

**Table 5** Accuracy (in %)

| Data set | SVM | ANN |
|----------|-----|-----|
| PC1 | 34.23 | **93.60** |
| PC2 | 13.67 | **97.72** |
| PC3 | 78.37 | **87.37** |
| PC4 | 82.70 | **90.67** |
| PC5 | 3.27 | **81.12** |
| KC2 | 0.95 | **94.50** |
| KC3 | 24.35 | **84.02** |

The values in bold shows the better performance of that particular technique/method as compared to the other methods/techniques in the tables.

**Table 6** Recall (in %)

| Data set | SVM | ANN |
|----------|-----|-----|
| PC1 | **47.06** | 45.45 |
| PC2 | **50.00** | 6.25 |
| PC3 | **8.62** | 0.00 |
| PC4 | 25.37 | **29.21** |
| PC5 | **59.81** | 31.39 |
| KC2 | 16.67 | **45.79** |
| KC3 | 11.47 | **33.33** |

The values in bold shows the better performance of that particular technique/method as compared to the other methods/techniques in the tables.

**Table 7** Specificity (in %)

| Data set | SVM | ANN |
|----------|-----|-----|
| PC1 | 33.33 | **98.46** |
| PC2 | 12.41 | **99.72** |
| PC3 | 89.24 | **99.78** |
| PC4 | 90.13 | **99.22** |
| PC5 | 1.41 | **99.58** |
| KC2 | 0.49 | **95.90** |
| KC3 | 70.59 | **95.56** |

The values in bold shows the better performance of that particular technique/method as compared to the other methods/techniques in the tables.

On the account of accuracy, ANN has clearly outperformed SVM for all the seven data sets with the maximum and minimum accuracy of 97.92 and 81.12%, respectively. SVM could not even achieve a satisfactory performance with a minimum accuracy of 0.95% and a maximum of 82.07%. However, when compared on recall or the probability of detection, SVM performed better than the artificial neural networks on four out of seven cases, with a maximum recall value of 59.81%. ANN also performed better than SVM with specificity as the evaluation measure in all the seven cases with a maximum value of 99.78%. However, recall is generally considered as an important parameter over others because it is essential to classify the defective modules correctly.

# 6 Conclusions

A comparative study of the traditional support vector machine and the novel artificial neural network classification models in the context of software defect prediction has been reported in the paper. The defect prediction models were built using the seven defect data sets from the PROMISE repository in the MATLAB environment. A linear kernel based SVM and a multilayer feed forward ANN was built. The models were empirically compared on the parameters of accuracy, recall and specificity. The techniques performed differently along different measures. ANN performed well on the dimensions of accuracy and specificity while SVM performed better when evaluated with recall. Hence, it is concluded that it is the criticality of the software project which helps to decide the measure to be used for performance evaluation and hence, determine the classification model. Also, it is encouraged to perform similar studies on a larger set of data sets to determine if the results obtained in this study can be generalized.

# References

1. Chidamber, S.R., Kemerer, C.F.: A metrics suite for object oriented design. IEEE Trans. Softw. Eng. **20**(6), 476–493 (1994)
2. Bailey, C.T., Dingee, W.L.: A software study using Halstead metrics. ACM SIGMETRICS Perform. Eval. Rev. **10**(1), 189–197 (1981)
3. Khoshgoftaar, T.M., Gao, K., Szabo, R.M.: An application of zero-inflated poisson regression for software fault prediction. In: Proceedings of 12th International Symposium on Software Reliability Engineering, pp. 66–73 (2001)
4. Challagulla, V.U.B., Bastani, F.B., Yen, I.-L., Paul, R.A.: Empirical assessment of machine learning based defect prediction techniques. Int. J. Artif. Intell. Tools **17**(2), 389–400 (2008)
5. Elish, K.O., Elish, M.O.: Predicting defect-prone software modules using support vector machines. J. Syst. Softw. **81**(5), 649–660 (2008)
6. Vapnik, V.: The Nature of Statistical Learning Theory. Springer-Verlag, New York (1995)
7. Brooks, R.A.: Intelligence without representation. Artif. Intell. **47**(1–3), 139–159 (1991)
8. Venkatachalam, A.R.: Software cost estimation using artificial neural networks. In: Proceedings of International Joint Conference on Neural Networks, pp. 987–990 (1993)
9. Heiat, A.: Comparison of artificial neural network and regression models for estimating software development effort. Inf. Softw. Technol. **44**(15), 911–922 (2002)
10. Thwin, M.M.T., Quah, T.-S.: Application of neural networks for software quality prediction using object-oriented metrics. J. Syst. Softw. **76**(2), 147–156 (2005)
11. So, S.S., Cha, S.D., Kwon, Y.R.: Empirical evaluation of a fuzzy logic-based software quality prediction model. Fuzzy Sets Syst. **127**(2), 199–208 (2002)

12. Menzies, T., Krishna, R., Pryor, D.: The promise repository of empirical software engineering data; http://openscience.us/repo. North Carolina State University, Department of Computer Science (2016)
13. Lanubile, F., Lonigro, A., Visaggio, G.: Comparing models for identifying fault-prone software components. In: Proceedings of 7th International Conference on Software Engineering and Knowledge Engineering, pp. 312–319 (1995)
14. Gondra, I.: Applying machine learning to software fault-proneness prediction. J. Syst. Softw. **81**(2), 186–195 (2008)
15. Zheng, J.: Cost-sensitive boosting neural networks for software defect prediction. Expert Syst. Appl. **37**(6), 4537–4543 (2010)
16. Malhotra, R.: Comparative analysis of statistical and machine learning methods for predicting faulty modules. Appl. Soft Comput. **21**, 286–297 (2014)
17. Abe, S.: Support Vector Machines for Pattern Classification. Springer, USA (2005)
18. Burges, C.J.C.: A tutorial on support vector machines for pattern recognition. Data Min. Knowl. Disc. **2**(2), 121–167 (1998)
19. Gun, S.: Support vector machines for classification and regression. Technical Report, University of Southampton (1998)
20. Gershenfeld, N.A., Weigend, A.S.: Time series prediction: forecasting the future and understanding the past. In: The Future of Time Series. Addison-Wesley, pp. 1–70 (1993)
21. Hagan, M.T., Menhaj, M.B.: Training feedforward networks with the Marquardt algorithm. IEEE Trans. Neural Netw. **5**(6), 989–993 (1994)
22. More, J.J.: The Levenberg-Marquardt algorithm: implementation and theory. In: Lecture Notes in Mathematics, Numerical Analysis. Springer, Berlin, pp. 105–116 (2006)
23. Stehman, S.V.: Selecting and interpreting measures of thematic classification accuracy. Remote Sens. Environ. **62**(1), 77–89 (1997)

# Random Selection of Crossover Operation with Mutation for Image Encryption—A New Approach

Jalesh Kumar and S. Nirmala

**Abstract** Securing multimedia information from attacks is a challenging task. There is a need of a versatile technique to convert the multimedia information into the unintelligent form to protect from vulnerability. A new approach based on random selection of crossover operation and mutation to secure the digital images is proposed. It comprises three different stages. In the first stage, three different crossover operations namely, one point, two points and uniform crossover operations of a genetic algorithm are implemented. Crossover points for one point and two points are generated on the basis of pseudo random number generator. For a selection of crossover operations, 4-bit linear feedback register is designed to generate the unpredictable sequence in the second stage. In the third stage, an encryption process is carried out on the input image based on the selected crossover operation. Further, mutation operation based on exclusive OR operation is performed to enhance the unpredictability. Experiments are conducted for different types of images with different crossover operation. The proposed work is analyzed in terms of entropy, peak signal to noise ratio, structural and feature similarity index. From the result obtained it is evident that the proposed approach increases the randomness and enhances the security for the digital images.

**Keywords** One-point crossover · Two points crossover · Uniform crossover · Genetic algorithm · Mutation

J. Kumar (✉) · S. Nirmala
Computer Science Department, Jawaharlal Nehru National College of Engineering,
Shivamogga 577201, Karnataka, India
e-mail: jalesh_k@yahoo.com

S. Nirmala
e-mail: nir_shiv_2002@yahoo.co.in

© Springer Nature Singapore Pte Ltd. 2018
R.K. Choudhary et al. (eds.), *Advanced Computing
and Communication Technologies*, Advances in Intelligent Systems
and Computing 562, https://doi.org/10.1007/978-981-10-4603-2_7

# 1 Introduction

In spite of the advancement in communication technologies that made the transmission of multimedia information faster, it had also led to the vulnerability of the information. Medical images, satellite images, document images or personal information in the form of images needs protection during transmission. Encrypting the information in the disguised form is the common strategy to protect such information. Even though standard encryption algorithms like DES, AES or RSA are available, the algorithms are more suitable for textual information. The structure of information in the form of images hold the properties like redundancy, correlation, and size which makes the exigent demand for the image encryption techniques. There are a number of techniques available in the literature to secure the information contents of the image. Some of the image security techniques are substitution and permutation [1], SCAN pattern [2], chaos based [3] and bit manipulation [4]. Evolutionary procedures are additionally picking up the significance in security. Cellular automata, genetic algorithms and neural network are gaining the importance in the information security. Kumar [5] proposes a methodology in a genetic algorithm with pseudorandom grouping to scramble the information stream. The components of such a methodology incorporate high information security. The idea of genetic calculations is utilized along with chaos theory for key generation. Encryption process with crossover operation and a pseudorandom sequence are discussed in [6]. But the discussed method is time consuming. Husainy [7] discussed a new image encryption technique based on genetic process. Mutation and crossover operations are used for encryption. Different vector lengths and number of crossover and mutation operations determine the security. A frequency domain method based on genetic algorithm is proposed in [8]. Magnitude and phase modification using differential evolution is discussed. Crossover points are selected based on linear feedback shift register. In [9], the cellular automata sequences are considered for key generation. Single-dimensional and two-dimensional cellular automata, reversible and irreversible sequences are considered for the key sequence. DNA sequences are considered for the image encryption [10]. Four nucleic acids of the DNA are used to generate the unpredictable key sequence. Techniques of artificial neural networks for securing the information contents are described in [11]. Weights used in the network acts as a random key sequence. The advantage of the techniques is sender need not send the entire key sequence.

It is observed from the survey that the evolutionary approaches are gaining importance in the security. Crossover operations of genetic algorithms are used to generate the key sequence. However, in all the technique one type of crossover operation is used. In most of the technique chaos methods are included with genetic operations to enhance the security. In the proposed work, three different crossover operations are randomly selected based on linear feedback shift register. Based on the random selection, crossover operation is applied along with the mutation to enhance the security of images.

## 2 Proposed Work

In this work, image encryption based on random selection of crossover operation with mutation operation of genetic algorithm is carried out. Three different crossover operations namely one-point, two point, and uniform crossover operation are designed. Based on selection, the unpredictable sequence is produced. 4-bit linear feedback shift register is used to select the type of crossover operation. The block diagram of the proposed work is shown in Fig. 1. Three stages considered in the proposed work are as follows.

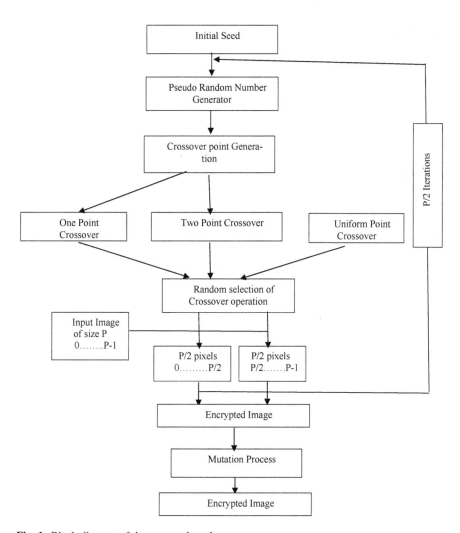

**Fig. 1** Block diagram of the proposed work

## *2.1   Generation of Different Types of Crossover Operation*

Pseudorandom number generator is initialized with 32-bit random number. Based on the number generated crossover point is considered. For one-point crossover, single value is selected and two numbers are selected from the sequence for two points crossover.

### 2.1.1   One-Point Crossover Operation

In one point crossover operation, one cross over point is considered among the parents to generate new two offsprings. Following example represents one-point crossover operation.

> Example :     Parent 1= 01010101     Parent 2 = 11001010
> At crossover point  4, new offspring generated  are
> Offspring 1 = 01011010     Offspring 2 = 11000101

### 2.1.2   Two Points Crossover Operation

Two crossover points are chosen among the parents to generate new two offspring in this process. Example shown below demonstrates the two points crossover operation.

> Example :    Parent 1= 01010101     Parent 2 = 11001010
> At crossover point 2 and 6, new offspring generated are
> Offspring 1 = 01001001   Offspring 2 = 11010110

### 2.1.3   Uniform Crossover Operation

In uniform crossover operation, crossover operation takes place uniformly among the parents to generate the new offspring.

> Example :    Parent 1= 01010101     Parent 2 = 11001010
> Offspring 1 = 01011001   Offspring 2 = 11000110

## 2.2 Selection of Crossover Points

Among the three crossover operations, one is selected randomly. For random selection, 4-bit linear feedback shift register is used. Tap sequence is selected on the basis of the primitive polynomial $x^4 + x + 1$. Two output bits from LSB are considered to generate the numbers from 0 to 3. Depending on the output sequence generated, crossover operation is chosen. '0' indicates one point crossover, '1' for two points crossover and '2' is uniform crossover operation. '3' indicates no crossover operation is selected.

## 2.3 Encryption of Image Based on Crossover and Mutation Operation

Input image '$I$' is decomposed into two parts '$I_1$' and '$I_2$'. Crossover operation is carried out between each pixels of '$I_1$' and '$I_2$'. Mutation operation is applied on output image obtained after crossover operation to get the encrypted image. For mutation, Exclusive OR operation is performed.

## 3   Result of the Proposed Work

Experimental study is conducted on the image corpus which consists of 100 images of different size. Corpus contains colored images, images with textual information and gray images. Some of the images are from usc.spc. database [12]. One point, two points and uniform crossover operations are applied separately on the input images. Result obtained after each crossover operations for the input image in Fig. 2a are shown in Fig. 2b–d, respectively. Figure 2e shows the encrypted image obtained after selecting the three crossover operations in random. For random selection of two crossover operations one point and two points is shown in Fig. 2f. Figure 2 shows the encrypted images obtained after performing mutation operation on the output images in Fig. 3.

From the results in Figs. 2 and 3, it reveals that the encrypted image using randomly selected three different crossover operations contains less residual information and unable to reveal the original information contents of the image. Figure 4 shows the output image after the decryption process, which indicates there is no loss of the information contents after decryption.

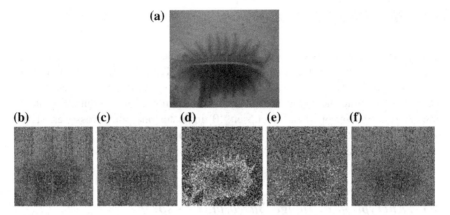

**Fig. 2** Encryption process for the sample image in the corpus. **a** Input image. **b** One-point crossover. **c** Two points crossover. **d** Uniform crossover. **e** Randomly selected one, two and uniform crossover. **f** Randomly selected one and two points crossover

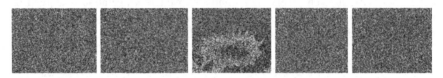

**Fig. 3** Mutation operation on the output images after crossover operation in Fig. 2

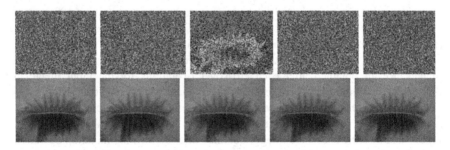

**Fig. 4** Decryption process

## 4   Security Analysis

Statistical measures are used to analyze the encrypted images. To evaluate the performance of the proposed method, statistical analyses are carried out in terms of structural similarity index, feature similarity index, entropy, and correlation coefficient values.

## 4.1 Structural Similarity Index

It is a measure of similarity between two images. The measure between two signals $x$ and $y$ is given by formula below [12]:

$$\text{SSIM} = \frac{(2\mu_x\mu_y + c_1)(2\sigma_{xy} + c_2)}{(\mu_x^2 + \mu_y^2 + c_1)(\sigma_x^2 + \sigma_y^2 + c_2)} \tag{1}$$

with

$\mu_x$  the mean of $x$ and
$\mu_y$  the mean of $y$
$\sigma_x^2$  the variance of $x$ and
$\sigma_y^2$  the variance of $y$
$\sigma_{xy}$  the covariance of $x$ and $y$

$c_1 = (k_1L)^2$ and $c_2 = (k_2L)^2$ are two variables to stabilize the division with weak denominator. Where '$L$' is the dynamic range of the pixel values (255 for 8-bit grayscale images), and $k_1 \ll 1$ and $k_2 \ll 1$ are small constants [12]. The value of SSIM is in the range [0 1]. The value '0' indicates that there is no correlation between two images. Value '1' indicates two images are similar.

## 4.2 Information Entropy

Entropy of the information is calculated as follows [13],

$$\text{Entropy} = \sum P(i) \log_2 \frac{1}{P(i)} \tag{2}$$

The probability of occurrence of a pixel with value '$i$' is $P(i)$.

## 4.3 Feature Similarity Index

On the basis of phase component and gradient magnitude feature extracted from two images feature similarity index is computed [14].

$$\text{FSIM} = \frac{\sum_{x\in\Omega} S_L(x) \cdot PC_m(x)}{\sum_{x\in\Omega} PC_m(x)} \tag{3}$$

$\Omega$   is whole image in spatial domain.

where $PC_m(x) = \text{Max}\ (PC_1(x), PC_2(x)), PC_1(x)$ and $PC_2(x)$ are phase component of two images.

$S_L(x) = [S_{PC}(x)]^\alpha \cdot [S_G(x)]^\beta$ where $S_{PC}(x)$ is similarity measure of phase component and $S_G(x)$ is similarity measure of gradient magnitude. $\alpha$ and $\beta$ are adjusting parameters. For simplicity $\alpha = \beta = 1$. FSIM values are in the range [0 1]. The value '1' indicates the features of two images are same. The value less than 1 indicates features are not similar.

## 4.4   Feature Similarity Index with Chromatic Information

Feature similarity index by considering chromatic information are computed using the Eq. 4 [14]. The value of FSIMc ranges from [0 1]. '1' indicates two images are similar with respect to chromatic information. The value '0' indicates there is no similarity between two images.

$$\text{FSIM}_c = \frac{\sum_{x \in \Omega} S_L(x) \cdot [S_c(x)]^\lambda \cdot PC_m(x)}{\sum_{x \in \Omega} PC_m(x)} \tag{4}$$

$\Omega$   is whole image in spatial domain.

where $\lambda > 0$ is the parameter used to adjust the importance of the chromatic components. $S_c(x)$ is chrominance similarity measure between two images.

The performance analysis is evaluated in terms of parameters PSNR [13], SSIM, entropy, FSIM, and FSIMc. Average value of these parameters for all images in the corpus is tabulated in Table 1. From the values tabulated in Table 1, it is evident that the random selection of one, two and uniform crossover operation exhibits high unpredictability. Compared to only one point, two points, and uniform crossover operation, in random selection of three crossover operation enhances the amount of randomness in encrypted image. Even though random selection of one and two points crossover operation increases the unpredictability, but combination of three crossover operation achieves improved performance.

**Table 1** Average value of the performance metrics

| Metrics | One point | | Two point | | Uniform | | Random one, two, and uniform | | Random one and two | |
|---|---|---|---|---|---|---|---|---|---|---|
| | Crossover | Mutation | Crossover | Mutation | Crossover | Mutation | Crossover | Mutation | Crossover | Mutation |
| PSNR | 14.16 | 10.77 | 13.84 | 10.83 | 12.07 | 11.34 | 12.54 | 10.79 | 13.86 | 10.79 |
| Entropy | 7.6 | 7.98 | 7.7 | 7.99 | 7.83 | 7.96 | 7.82 | 7.994 | 7.67 | 0.7991 |
| SSIM | 0.10 | 0.0212 | 0.085 | 0.0213 | 0.193 | 0.133 | 0.0558 | 0.0211 | 0.086 | 0.0215 |
| FSIM | 0.47 | 0.514 | 0.459 | 0.516 | 0.559 | 0.638 | 0.418 | 0.514 | 0.46 | 0.513 |
| FSIMc | 0.45 | 0.488 | 0.436 | 0.49 | 0.53 | 0.609 | 0.396 | 0.488 | 0.43 | 0.488 |

## 5 Conclusions

In this work, three different crossover operations of genetic algorithm have been analyzed. Based on the combinations of one point, two points, and uniform crossover operations, a security model is implemented. To enhance the security, mutation operation is applied after crossover operation. It has been observed from the experimental results that the proposed approach increases the unpredictability and further improves the security.

## References

1. El-Wahed, M.A., et al.: Efficiency and security of some image encryption algorithms. In: Proceedings of World Congress on Engineering, vol. 1, pp. 822–1706 (2008)
2. Panduranga, H.T., Naveen Kumar, S.K.: Hybrid approach for image encryption using scan patterns and carrier images. Int. J. Comput. Sci. Eng. 2(2) (2010)
3. Usama, M., Khan, M.K., Alghathbar, K., Lee, C.: Chaos-based secure satellite imagery cryptosystem. Comput. Math Appl. 60(2), 326–337 (2010)
4. Parameshachari, B.D., Soyjaudah, K.S., Sumithra Devi, K.A.: Image quality assessment for partial encryption using-modified cyclic bit manipulation. Int. J. Innov. Technol. Explor. Eng. (2013)
5. Kumar, A., Ghose, M.K.: Overview of information security using genetic algorithm and chaos. Inf. Secur. J.: A Glob. Perspect. 18, 306–315 (2009)
6. Kumar, A., Rajpal, N.: Application of genetic algorithm in the field of steganography. J. Inf. Technol. 2(1), 12–15 (2004)
7. Husainy, M.: Image encryption using genetic algorithm. Inf. Technol. J. 5(3), 516–519 (2006)
8. Abuhaiba, I.S.I., Hassan, M.A.S.: Image encryption using differential evolution approach in frequency domain. Signal Image Process. Int. J. (SIPIJ). 2(1) (2011)
9. Das, d., Ray, A.: A parallel encryption algorithm for block ciphers based on reversible programmable cellular automata. Arxiv preprint arXiv:1006.2822 (2010)
10. Zhang, Q., Xue, X., Wei, X.: A novel image encryption algorithm based on dna subsequence operation. Sci. World J. (2012)
11. Kanter, I., Kinzel, W.: The theory of neural networks and cryptography. Quant. Comput. Comput. 5(1), 130–140 (2005)
12. Wang, Z., Bovik, A.C., Sheikh, H.R., Simoncelli, E.P.: Image quality assessment: from error visibility to structural similarity. IEEE Trans. Image Process. 13(4), 600–612 (2004)
13. Kumar, J., Nirmala, S.: A new light weight encryption approach to secure the contents of image. In: International Conference on Advances in Computing, Communications and Informatics ICACCI, 2014, pp. 1309–1315. IEEE (2014)
14. Zhang, L., Zhang, L., Mou, X., Zhang, D.: FSIM: a feature similarity index for image quality assessment. IEEE Trans. Image Process. 20(8), 2378–2386 (2011)

# User-Interactive Recommender System for Electronic Products Using Fuzzy Numbers

**Shalini Sharma, C. Rama Krishna, Shano Solanki, Khushleen Kaur and Sharandeep Kaur**

**Abstract** Recommender systems support users in decision-making processes like shopping, entertainment, browsing and reading from the extensive information available on the Internet. A customer often feels difficulty in expressing exact requirements when confronted with the purchasing of a high-tech product having complex characteristics, e.g., digital camera, smartphone, notebook, electronic tablet, kindle, server, personal computer, car, etc. Most of the consumers are not so much aware with the technical features of these electronic products. As these products are expensive and users are not going to buy these items regularly so the system is not aware with user past profile, purchases and interests, on the basis of which it can generate recommendations; this is a kind of cold start problem. In this paper, consumer's requirements are obtained from the human–computer interaction based on two kinds of fuzzy numbers, viz., Trapezoidal and Triangular. This is also evaluated which fuzzy number method gives better results. The field of proposed recommender system is digital camera area.

**Keywords** Euclidean fuzzy distance · Coefficient of correlation · Recommender system · Fuzzy numbers

S. Sharma (✉) · C.R. Krishna · S. Solanki · K. Kaur · S. Kaur
Department of Computer Science & Engineering, National Institute of Technical Teachers'
Training and Research, Sector-26, Chandigarh 160019, India
e-mail: shalu0502@gmail.com

C.R. Krishna
e-mail: ramakrishna.challa@gmail.com

S. Solanki
e-mail: s_solanki_2000@yahoo.com

K. Kaur
e-mail: khushleendhanoa@gmail.com

S. Kaur
e-mail: sharandeep.nitttr@gmail.com

© Springer Nature Singapore Pte Ltd. 2018
R.K. Choudhary et al. (eds.), *Advanced Computing
and Communication Technologies*, Advances in Intelligent Systems
and Computing 562, https://doi.org/10.1007/978-981-10-4603-2_8

# 1  Introduction

Recommender Systems (RS) are application software techniques suggesting items to a user [1] and filtering and sorting items and information [2]. The extensive information available on the Internet and the frequent launch of new e-business services have confused users and has often lead them to wrong decisions. So, due to vast available choices, the number of users has decreased, thereby reducing the profits. RS are valuable for users to deal with the overloaded data over the internet and are very capable, powerful and dominant tools in e-business.

Customers buying high-technology products like laptops, desktops, digital camera, smartphone, and car with many technical features; face difficulty to specify their requirements because most of consumers often do not know the technical features of the products. A common consumer purchases these products less frequently, so enterprises have lack of sufficient information about the customer's previous purchases, so it is impossible to use a customer's prior preferences. The above case is a kind of cold start problem. Therefore, there is a need for developing a user interactive RS, where the users should only input their requirements as desired product functionality. This paper proposes a way to find requirement—product similarity for generating appropriate recommendation. The proposed system is developed into digital camera area.

The remaining paper is organized as follows. Section 2 explains related work about RS that is built upon functional needs and a product's technical features. Section 3 presents our proposed work for product recommendation in detail. Section 4 reports about evaluations and results. Finally, we equip conclusion in Sect. 5.

# 2  Related Work

After reviewing the literature, it is found that RS is emerging technology to support decision-making of users. In particular, a lot of research has been done on various RS techniques, like content based filtering [2, 3], collaborative filtering technique [4, 5], hybrid recommendation approaches [6–8], knowledge-based RS [9–15]. In the literature, researchers have proposed solutions for various recommender system challenges like data sparsity problem [16], cold start problem [17, 18]. All of the techniques are based on past records of the user like user profile, rating of user for a particular product. There is no restriction on user to describe his needs explicitly. In knowledge-based RS, the system obtains consumers' needs for a item, and then uses that data as the base to elect the most appropriate products for the user [10]. Nonetheless, in most of these RS, user input their needs in terms of technical features of the products. But consumers do not know the typical technical features of sophisticated high-tech products. To solve this problem, Hendratmo and Baizal [19] presented a plan to develop a conversational RS that is capable to fetch the practical needs of the consumer by a question–answer mechanism. Meantime, Cao and Li [1] developed an impressive model to encompass user's functional needs in a RS. Baizal and Adiwijaya [20] extended the work of Cao and Li for Smartphones. In this scenario recent work has been done on triangular fuzzy numbers. We proposed a system that will recommend items both for triangular and trapezoidal fuzzy numbers and compare which type of numbers give better results.

# 3 System Architecture

The system explicitly inquires users' functional requirements through some questions. For each requirement, the user has to input only degree of interest (DOI) in fuzzy scale, such as very low (VL), low (L), medium (M), high (H) and very high (VH). Interest level of user about a functional need is represented by the DOI. Main processes of the system are as follows:

1. Acquiring user's category as beginner, family, traveler, enthusiast, or sportsperson and DOI for functional requirements through some of questions given by the system. The answers of user will then be passed on for mapping the needs to digital camera characteristics.
2. Obtaining a quality level of each feature of product from weight value of digital camera component.
3. Recommending digital camera to user using two types of fuzzy numbers. The similarity level of needs is compared to the presently marketed digital camera using Euclidean fuzzy distance.
4. Comparing results of the two fuzzy numbers with the database of best cameras for particular category using Spearman's rank correlation coefficient and delivering the better method.

## 3.1 Fuzzy Numbers

Fuzzy numbers allow composing the mathematical model of linguistic variable. Value of a fuzzy number is imprecise, rather than exact. Fuzzy number means a real number interval with fuzzy boundary.

*Triangular fuzzy number* (*TFN*): It is defined by a triplet ($a1$, $a2$, $a3$). The membership function is described as Eq. (1) [21].

$$u(x) = \begin{cases} (x - a1)/(a2 - a1) & a1 \leq x \leq a2 \\ (a3 - x)/(a3 - a2) & a2 \leq x \leq a3 \\ 0 & \text{otherwise} \end{cases} \quad (1)$$

Linguistic terms are used to help users easily express their interest and the system can easily access product features. Five linguistic sets are allowable: (1) Very Low, (2) Low, (3) Medium, (4) High, (5) Very High.

*Trapezoidal fuzzy number (TrFN)*: It is a fuzzy number having quadruplet ($a1$, $a2$, $a3$, $a4$) and its membership function is described as Eq. (2) [21].

$$u(x) = \begin{cases} (x - a1)/(a2 - a1) & a1 \leq x \leq a2 \\ 1 & a2 \leq x \leq a3 \\ (x - a4)/(a3 - a4) & a3 \leq x \leq a4 \\ 0 & \text{otherwise} \end{cases} \tag{2}$$

## 3.2 Euclidean Fuzzy Distance

It is a fuzzy method that considers range and refers to measurement of the compactness between two membership functions. The study determines similarity of user's requirements and product's components, using modified Euclidean fuzzy distance equation Eq. (3) introduced by Baizal and Adiwijaya [20].

$$N_E(\widetilde{q}_A, \widetilde{q}_B) = 1 + \frac{1}{\sqrt{3}} \left( \sum_{j=1}^{3} |q_A^j - q_B^j|^2 \right)^{1/2} \tag{3}$$

where in the case of triangular fuzzy number $\widetilde{q}_A = (q_A^1, q_A^2, q_A^3)$ is compared TFN, $\widetilde{q}_B = (q_B^1, q_B^2, q_B^3)$ is target TFN. In the case of trapezoidal fuzzy number $\widetilde{q}_A = (q_A^1, q_A^2, q_A^3, q_A^4)$ is compared TrFN, $\widetilde{q}_B = (q_B^1, q_B^2, q_B^3, q_B^4)$ is target TrFN. The $N_E$ will be processed $n$ times for top-$n$ recommendations. If the resultant number is near to 1, the similarity between membership triplets/quadruplets is more. Eq. (4) gives the similarity between two fuzzy sets [1].

$$N_E(\widetilde{X}, \widetilde{Y}) = \sum_{i=1}^{n} \left( N_E(\widetilde{x}_i, \widetilde{y}_i) \times v_i \right) \tag{4}$$

where $N_E(\widetilde{x}_i, \widetilde{y}_i)$ is identical to that in Eq. (1) and $N_E(\widetilde{X}, \widetilde{Y})$ is the closeness between fuzzy sets $\widetilde{X}$ and $\widetilde{Y}$, and $v_i$ is the correspondent weight to the ith TFN/TrFN, where

$$\sum_{i=1}^{n} v_i = 1 \tag{5}$$

Eq. (6) gives the component capability [1].

$$p_i^k = \sum_{j=1}^{n} \left( f_i^{jk} \times w_i^j \right) \qquad (6)$$

where p is component capability of camera, $k$ stands for $k^{th}$ TFN, $i$ refer to $i^{th}$ component, $j$ refers to $j^{th}$ feature of $i^{th}$ component. From Eq. (7) the most ideal component for the user requirement is achieved [1].

$$c_i^k = \sum_{j=1}^{n} \left( u_i^{jk} \times v_i^j \right) \qquad (7)$$

From Eq. (7), we can observe that to get the value of the component ideal for the consumers, the computation of the requirement of user is mapped to component (u) and multiplied to the weight of the mapping of users' requirement toward the component of digital camera (v). It will be a valid input for Euclidean Fuzzy, if the total of the weight is 1.

## 3.3 Attaining Matching Components to User's Functional Requirements

The system asks questions on need of user which may assuredly help users who are not usual with technological features of product. The user interface asks functional requirements of user through some questions. Each functional requirement will be mapped to the corresponding components based on DOI value. Table 1 describes the weights given to functional requirement towards digital cameras component.

Table 1 The relationship between the critical component and customer qualitative need

| Customer qualitative need | Ability weight (v) | Components |
|---|---|---|
| Photo quality | 0.1 | Lens |
| | 0.4 | Aperture |
| | 0.5 | Image sensor |
| Price | 0.1 | Aperture |
| | 0.2 | Image sensor |
| | 0.28 | Memory card |
| | 0.12 | LCD screen |
| | 0.1 | Flash |
| | 0.1 | User controls |
| | 0.1 | Angle of view |
| ⋮ | ⋮ | ⋮ |

**Table 2** Technical features of critical component in a digital camera

| Components | Feature weight ($w_i^j$) | Technical feature | Candidates |
|---|---|---|---|
| Lens | 0.1 | Focal length | 10, 50, 85 mm, etc. |
| | 0.4 | Lens mount | Canon EF-M, Canon EF-S |
| | 0.5 | Lens F-ratio | f/1.4, f/4.5, f/5, f/6.3, f/8 etc. |
| Shutter release | 1 | Shutter speed | faster than 1 s (1/125); slower than 1 s (1″) |
| Aperture | 0.25 | Aperture sizes | f/1.4, f/22, etc. |
| | 0.25 | Aperture area | 490.8, 122.7 mm² etc. |
| | 0.25 | Aperture speed | exposure time; 200 ISO, 400 ISO |
| | 0.25 | Aperture range | 35, 24–200 mm etc. |
| Memory card | 0.6 | Size | 16, 32, 64 GB etc. |
| | 0.4 | Speed capacity | 30, 10 MB/s |
| Viewfinder | 1 | Type | Waist-level (reflecting); sports viewfinder; twin and single-lens reflex |
| ⋮ | ⋮ | ⋮ | ⋮ |

## 3.4 Component's Quality Level Based on Features

Based on existing mapping, the system will classify the quality features and then it will be input to Euclidean Fuzzy distance. The system will calculate the component quality of digital camera using Eq. (1). This mapping of product features-components are illustrated in Table 2.

## 4 The Evaluations

The evaluation is done by comparing the generated recommendations with top best cameras categorized according to the type of the user. Recommendations generated using both methods are compared separately with the corresponding database and Spearman's coefficient is calculated for both of these methods. The method for which the value of Spearman coefficient is more close to 1; is better method.

## 4.1 Recommendations Generation

Python language is used for the implementation of the work in Anaconda2-4.1.1 Spyder environment. Results for both methods are displayed separately.

## 4.2   Comparison Result for Two Methods

Comparison of result is done for the same user input in both evaluation situations using Spearman's coefficient according to category of the user. The $n$ rows scores $X_i$, $Y_i$ for a sample of size $n$ are transformed to ranks $x_i$, $y_i$ and correlation $\rho$ is computed using Eq. (8).

$$\rho = 1 - \frac{6 \sum d_i^2}{n(n^2 - 1)} \quad (8)$$

where $d_i = x_i - y_i$ is the difference between ranks. The results are shown in Fig. 1. For example, let us consider that user selected type as Enthusiast and entered the preferences and recommendations were generated according to the preferences. These recommendations are stored in a table. One instance of this scenario is

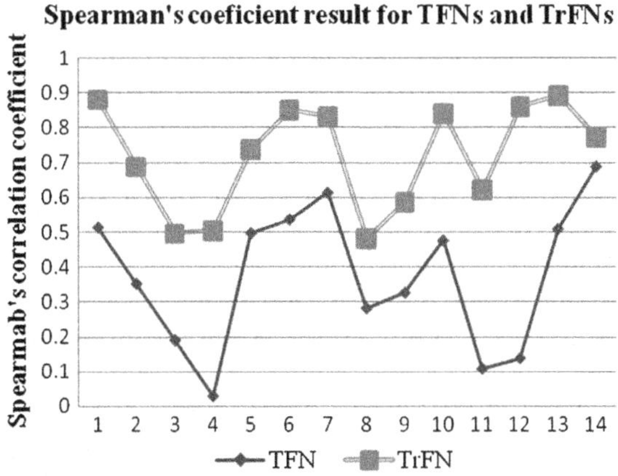

**Fig. 1** Results

Spearman's coeficient result for TFNs and TrFNs

**Table 3** Recommended cameras to the user using TFNs and TrFNs

| Rank | Recommendation for TFNs | Recommendation for TrFNs |
|---|---|---|
| 1 | Nikon COOLPIX AW120 16 MP Wi-Fi and Waterproof Digital Camera | Nikon COOLPIX AW120 16 MP Wi-Fi and Waterproof Digital |
| 2 | Nikon D90 DX-Format CMOS DSLR Camera | Canon EOS Rebel SL1 Digital SLR |
| 3 | Nikon D5500 | Nikon D5 |
| 4 | Panasonic Lumix DMC-ZS100 | Canon PowerShot G7 X Mark II |
| 5 | Canon PowerShot G9 X | Sony Alpha ILCE-A6000 |
| 6 | Sony Alpha ILCE-A6000 | Sony Cyber-shot DSC-RX10 III |
| 7 | Canon PowerShot G7 X Mark II | Canon PowerShot G9 X |
| 8 | Sony Cyber-shot DSC-RX10 III | Canon EOS M3 |
| 9 | Canon EOS M3 | Canon EOS 5D Mark III |
| 10 | Canon EOS 80D | Panasonic Lumix DMC-ZS100 |

**Table 4** Best cameras for enthusiast

| Rank | Camera model |
|------|--------------|
| 1 | Canon PowerShot G7 X Mark II |
| 2 | Sony Cyber-shot DSC-RX10 III |
| 3 | Sony Alpha ILCE-A6300 |
| 4 | Canon EOS 80D |
| 5 | Canon PowerShot G9 X |
| 6 | Sony Alpha ILCE-A6000 |
| 7 | Sony Cyber-shot DSC-RX100 III |
| 8 | Canon EOS 5D Mark III |
| 9 | Canon EOS M3 |
| 10 | Panasonic Lumix DMC-ZS100 |

presented in Table 3 for TFNs and TrFNs. This table is compared to the best cameras shown in Table 4 [22] for Enthusiast category according to the ranks.

Spearman's coefficient for TFNs is: 0.2303030 and for TrFNs is: 0.769696969. Like this, we run the system 80 times for testing. Score for Spearman's correlation for two methods were stored for each run. The resultant graph for Spearman's correlation values for TFNs and TrFNs for 15 runs is shown in the following subplot in Fig. 1.

# 5 Conclusion

In conclusion, it may be argued that both the methods are effective in this scenario. But in terms of better recommendation, it can be said that trapezoidal fuzzy method performs better as compared to triangular fuzzy methods. Though recommendations are generated through many techniques but for user interactive RS fuzzy methods have been proved beneficial. A prototype for the RS is developed with 350 camera dataset and in 92% cases trapezoidal fuzzy numbers proved better than triangular fuzzy numbers. From these results, it can be wrapped up that trapezoidal fuzzy-based method is best for recommending better products in this scenario.

# References

1. Cao, Y., Li, Y.: An intelligent fuzzy-based recommendation system for consumer electronic products. Expert Syst. Appl. **19**, 230–240 (2007)
2. Shahabi, C., Banaei-Kashani, F., Chen, Y., McLeod, D.: Yoda: an accurate and scalable web-based recommendation system. In: Batini, C. (ed.) Cooperative Information Systems, 1st edn, pp. 418–432. Springer, Berlin (2001)
3. Shardanand, U., Maes, P.: Social information filtering: algorithms for automating "word of mouth". In: Proceedings SIGCHI Conference on Human Factors in Computing Systems. ACM Press/Addison-Wesley, New York, pp. 210–217 (1995)

4. Musiał, K., Juszczyszyn, K., Kazienko, P.: Ontology-based recommendation in multimedia sharing systems. Int. J. Syst. Sci. **34**(1), 97–106 (2008)
5. Ricci, F., Nhat Nguyen, Q., Averjanova, O.: Exploiting a map-based interface in conversational recommender systems for mobile travelers. In: Sharda, N. (ed) Tourism Informatics: Visual Travel Recommender Systems, Social Communities, and User Interface Design, 1st edn. Information Science Reference, Hershey, New York, pp. 354 (2009)
6. Hsieh, S.M., Huang, S.J., Hsu C.C., Chang H.C.: Personal document recommendation system based on data mining techniques. In: Proceedings IEEE/WIC/ACM International Conference on Web Intelligence, Place? IEEE, pp. 51–57 (2004)
7. Kazienko, P., Kolodziejski, P.: Personalized integration of recommendation methods for E-commerce. Int. J. Comput. Sci. Appl. **3**, 12–26 (2016)
8. Ullah, F., Sarwar, G., Chang Lee, S., Deok Moon, K., Tae Kim, J., Kyung Park, Y.: Hybrid recommender system with temporal information. In: Proceedings International Conference on Information Networking (ICOIN). IEEE, pp. 421–425 (2012)
9. Felfernig, A., Isak, K., Szabo, K., Zachar, P.: The VITA financial services sales support environment. In: Proceedings 19th National Conference on Innovative Applications of Artificial Intelligence. AAAI Press, Vancouver, British Columbia, Canada, pp. 1692–1699 (2007)
10. Felfernig, A., Burke, R.: Constraint-based recommender systems: technologies and research issues. In: Proceedings 10th International Conference on Electronic Commerce (2008)
11. Mirzadeh, N., Bansal, M., Ricci, F.: Feature selection methods for conversational recommender systems. In: Proceedings International Conference on e-Technology, e-Commerce e-Service. IEEE, pp. 772–777 (2005)
12. Ricci, F., Nhat Nguyen, Q.: Acquiring and revising preferences in a critique-based mobile recommender system. IEEE Intell. Syst. **22**, 22–29 (2007). doi:10.1109/MIS.2007.43
13. Smyth, B., Reilly, J., McGinty, L., McCarthy, K.: Compound critiques for conversational recommender systems. In: Proceedings IEEE/WIC/ACM International Conference on Web Intelligence. IEEE, pp. 145–151 (2004)
14. Thompson, C.A., Goker, M.H., Langley, P.: A personalized system for conversational recommendations. J. Artif. Intell. Res. **21**, 393–428 (2004)
15. Chena, R., Bau, C., Chen, S., Huangb, Y.: A recommendation system based on domain ontology and SWRL for anti-diabetic drugs selection. Expert Syst. Appl. **39**, 3995–4006 (2012)
16. Chen, T., ku, T.: Importance-assessing method with fuzzy number-valued fuzzy measures and discussions on TFNs and TrFNs. Int. J. Fuzzy Syst. **10**, 92–103 (2008)
17. Chen, S., Wang, C.: House selection using fuzzy distance of trapezoidal fuzzy numbers. In: Proceedings International Conference on Machine Learning and Cybernetics. IEEE, pp. 1409–1411 (2007)
18. Ricci, F., Rokach, L., Shapira, B.: Introduction to recommender systems handbook, 1st edn. pp. 1–35 (2010)
19. Baizal, Z., Widyantoro, D.: A framework of conversational recommender system based on user functional requirements. In: Proceedings 2nd International Conference on Information and Communication Technology (ICoICT). IEEE, pp. 160–165 (2014)
20. Arnett, D., Baizal, Z.: Recommender system based on user functional requirements using Euclidean fuzzy. In: Proceedings 3rd International Conference on Information and Communication Technology (ICoICT). IEEE, pp. 455–460 (2015)
21. Isabels, K., Uthra, G.: An application of linguistic variables in assignment problem with fuzzy costs. Int. J. Comput. Eng. Res. (2012)
22. http://www.imaging-resource.com/WB/WB.HTM

# Selection of Genes Mediating Human Leukemia, Using Boltzmann Machine

Sougata Sheet, Anupam Ghosh and Sudhindu Bikash Mandal

**Abstract** The Boltzmann machine model for identification of some possible genes mediating different disease has been reported in this paper. The procedure involves grouping of gene-based correlation coefficient using gene expression data sets. The usefulness of the procedure has been demonstrated using human leukemia gene expression data set. The vying of the procedure has been established using three existing gene selection methods like Significance Analysis of Microarray (SAM), Support Vector Machine (SVM), and Signal-to-Noise Ratio (SNR). We have performed biochemical pathway, $p$-value, $t$-test, sensitivity, expression profile plots for identifying biological and statistically pertinent gene sets. In this procedure, we have found more number of true positive genes compared to other existing methods.

**Keywords** Boltzmann machine · $p$-value · $t$-test

## 1 Introduction

Cancer is a group of 100 diseases for which a group of cells undergoes insensitive outgrowth [1]. It causes wreck of contiguous tissues and sometimes expansion to other location in the body via blood. People are affected by cancer at all stage and risk for most varieties increase with age. In United States 54,000 new cases occurs yearly and 24,000 deaths due to cancer. Nearly 27,000 adults and 2000 children are

S. Sheet (✉) · S.B. Mandal
A.K. Choudhury School of Information Technology,
University of Calcutta, Kolkata, India
e-mail: sougata.sheet@gmail.com

A. Ghosh
Department of Computer Science and Engineering,
Netaji Subhash Engineering College, Kolkata, India
e-mail: anupam.ghosh@rediffmail.com

© Springer Nature Singapore Pte Ltd. 2018
R.K. Choudhary et al. (eds.), *Advanced Computing
and Communication Technologies*, Advances in Intelligent Systems
and Computing 562, https://doi.org/10.1007/978-981-10-4603-2_9

83

affected each year [2]. Different types of research efforts, including ones based on surgery, chemotherapy, radiotherapy are being made to fight against cancer.

In the field of microarray data analysis, the most difficult remittance is genes selection [3]. Gene expression data generally contains large number of variable genes compared to the number of samples. The customary data mining approach cannot be immediately used to the data due to this individuality problem. The analysis of gene expression data used dimension reduction methodology for this purpose.

We introduce a method based on Boltzmann models for identifying genes mediating normal and disease genes [4, 5] in this paper. We denote those models as Boltzmann Machine Model-1 (BMM-1) and Boltzmann Machine Model-2 (BMM-2). We select the most important group but at first we using correlation coefficient and form of genes. The procedure is applicable in this data-rich situation, i.e., if the number of samples is gigantic and compared to the dimension of all samples. The number of samples is low which is compared to the number of genes in this problem. From given microarray gene expression data sets we have produce more data and solved this problem.

The usefulness of the procedure, along with its outstanding result over several others procedure, has been demonstrated one microarray gene expression data set with cancer related to human leukaemia. The results have been compared three existing methods like Significance Analysis of Microarray (SAM) [6], Support Vector Machine (SVM) [7, 8] and Signal-to-Noise Ratio (SNR) [9]. The compared of the result has been made using $t$-test and $p$-value.

## 2  Methodology

Let us assume a set $G = (g_1; g_2; ....; g_n)$ of $n$ genes for every of which the first $m$ expression values in normal samples and the subsequent n expression values in diseased samples are known. Now we calculate the interrelation coefficient within twins of these genes based on their expression values in normal samples. Thus, the interrelation coefficient $R_{pq}$ within $p$th and $q$th genes is given by [10]

$$R_{pq} = \frac{\sum_{k=1}^{m} (g_{pk} - y_p) * (g_{qk} - y_q)}{\sqrt{\left(\sum_{k=1}^{m} (g_{pk} - y_p)^2\right)} * \sqrt{\left(\sum_{k=1}^{m} (g_{qk} - y_q)^2\right)}} \tag{1}$$

Here $y_p$ and $y_q$ are the mean of expression values of $p$th and $q$th genes, respectively, above normal samples. The interrelation coefficient assumes values in the interval $[-1, 1]$. When $R_{pq} = -1$ (+1), there is a strong negative (positive) interrelation between $p$th and $q$th genes. Interrelation values which are high positive are placed into the same group. The main idea of grouping is as follows. Now interrelations with another gene if a gene has a strong positive, then the expression patterns of these two genes are similar. We may consider one of them as a typical gene and ignore the other genes in that situation. The genes in the same group are strongly positively correlated if the group of genes is identified in such a way. In order to do this, $R_{pq}$ (Eq. 1) is computed for each pair of genes. The genes are located in the similar group if $R_{pq} \geq 0.50$. Now we have applied interrelation coefficient to narrow downhearted the invention space by searching genes of a comparable behavior in terms of related expression patterns. The set of responsible genes mediating certain cancers are identified in this procedure. The choice of 0.50 as a threshold value has been done through extensive experimentation for which the distances among the cluster center have become maximize. The first group of genes is obtained in this way.

The Boltzmann machine is designed to consent useful searches for combination of hypotheses that is stored constraints and satisfy some input data [11]. A binary state whose two states are represented as the authentic values of the hypothesis is represented in every hypothesis. A content-addressable memory can be finalized by using distributed patterns of large combinations of hypotheses to continue for the kinds of difficult items for which we have words. New items are stored by varying the interactions between units so as to generate new permanent patterns of activity, and they are improving by establishment into the pattern of movement under the impact of an exotic input vector which is acts as a partial statement of the basic item. A good way to approach the excellent-fit problem is to define measure of how poorly the present pattern of activity in a module fits the external input and internal constraints and then we create the separate hardware units and minimize this measure. For this reason, we can create separate hardware units act so as to minimize this measure. The energy measure can be calculated with states of a binary network, and we generalize this measure to involve maintained inputs from outside the network which is shown by Hopfield network. Fully connected Boltzmann machine shown in Fig. 1.

$$\varepsilon = 1/2 \sum_{mn} w_{mn} \sigma_m \sigma_n - \sum_m (\mu_m - \beta_m) \sigma_m \qquad (2)$$

**Fig. 1** Fully connected Boltzmann machine

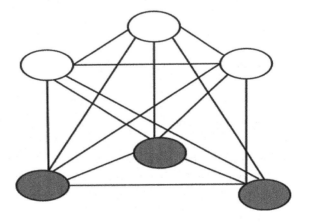

Now we can discover a local energy minimum in this kind of network is to constantly change all unit into whichever of its two states output the lower total energy given the current states of the another units. The procedure can minimize the energy, if hardware units form their decisions at haphazard, asynchronous moment and if sending times are ignorable so that each unit always "sees" the preset states of the units, and the network must resolve into an energy minimum. If all the link power is symmetric, which is classically the case for constraint amusement problems; every unit can enumerate its outcome on the total energy from information that is locally available. The distinction between the energy with the $k$th unit false and with it true is just

$$\Delta \varepsilon_k = \sum_m w_{km} \sigma_m + \mu_k - \beta_k \tag{3}$$

So the rule for reducing the total energy is to accept the true state, if the joined of external and internal input to the unit exceeds its threshold. This is called the familiar rule for binary threshold units. It is likely to exclude from poor local minimum and searches excellent ones by changing the simple procedure to allow occasional jumps to states of higher energy. At first sight, this point looks like a dirty hack which can never surety that the global minimum will be found. However, his entire module will treat in an effective pathway that can be analyze using statistical procedure that provided all units adopts the state with a probability given by

$$\pi_k = \frac{1}{1 + \chi^- \Delta \varepsilon_k / \tau} \tag{4}$$

where $\tau$ is a scaling parameter that acts like the temperature of a physical system. This procedure, which is similar the input–output function for a cortical neuron confirm that when the system has reached "thermal equilibrium" the relative probability of searching it in two global states is a Boltzmann distribution and is therefore determinate only by their energy difference

$$\frac{\pi_\gamma}{\pi_\delta} = \chi^{-(\varepsilon_\gamma - \varepsilon_\gamma)} / \tau \tag{5}$$

If $\tau$ is big equilibrium is extended quickly but the bias in support of the minor energy states is small. If $\tau$ is small, the bias is helpful but the time necessary to reach equilibrium is long.

## 3  Analysis of the Result

In this work, we can select one type of data set. The name of the data set is Waldenstrom's macroglobulinemia (B lymphocytes and plasma cells). It has been applied for the solution of B lymphocytes (BL) and plasma cells (PC) from patients

**Table 1** Comparative result on number of attributes of various sets of genes

| Data set | Gene set | BMM-1 | BMM-2 | SAM | SVM | SNR |
|---|---|---|---|---|---|---|
| Leukemia expression data | First 5 | 85 | 82 | 60 | 79 | 17 |
| | First 10 | 91 | 87 | 62 | 88 | 27 |
| | First 15 | 104 | 100 | 71 | 97 | 37 |
| | First 20 | 112 | 101 | 80 | 106 | 42 |

with Waldenstrom's macroglobulinemia (WM). The data set ID is GDS-2643 [12]. The total data set consists of 22,283 numbers of genes with 56 samples. Among them, there are 13 normal samples which consist of 8 normal for B lymphocytes and 5 normal plasma cells and 43 diseased samples which consist of 20 Waldenstrom's macroglobulinemia, 11 chronic lymphocytic leukemia, 12 multiple myeloma samples. The database web link is http://ncbi.nlm.nih.gov/projects/geo/.

In this section, the usefulness of the procedure is demonstrated on human leukemia gene expression data set and also included a comparative analysis with SAM, SVM and SNR. We have got 8 groups containing 2741, 2856, 2691, 2476, 2786, 2813, 2784, 2673 number of genes respectively. In this group, 2856 number of genes has been selected as most important group by both BMM-1 and BMM-2. Given in Table 1.

We have been determinate to compare the result of this leukemia gene expression data. In order to validate the results statistically, we have performed $t$-test on the genes identified by BMM-1 and BMM-2 on each data sets. $t$-test is the statistical significance which indicates whether or not the difference between two groups average most likely reflects an original difference in the population from which the group wear sampled. The $t$-value show the most significant genes (99.9%) which $p$-value <0.001. For these three types of data sets, we can apply $t$-test and we get corresponding $t$-value. We have identify some important genes like IARS (5.98), MMP25 (4.58), TYMS (3.96), HPS6 (5.59), MLX (5.32), CALCA (4.12), HIC2 (5.02), ANP32B (4.56), TFPI (5.72), CRYAB (3.98), NCF1C (3.39), HNRNPH1 (4.92), etc.

The numbers in the bracket show $t$-value of the corresponding gene. The $t$-value of this genes exceeds the value for $P = 0.001$. This means that this gene is highly significant (99.9% level of significance). Similarly genes like ERCC5 (3.12), PRDM2 (3.17), PRIM2 (2.61), TPT1 (3.29), RPS26 (2.83), EFCAB11 (3.22), PRPSAP2 (3.57), PRKACA (2.84), etc., exceed the $t$-value for $p = 0.01$. It indicates that these genes are significant at the level of 99%. Similarly genes like MED17 (2.34), MAPK1 (2.42), PIK3CB (2.05), NMD3 (2.34), ARG2 (2.19), EXOC3 (2.16), WHSC1 (2.18), RFC4 (2.26), GLB1L (2.41), HNF1A (2.05) etc., exceeds the value for $p = 0.05$. It indicate that this genes significant at the level of 95%. Similarly genes like FLG (1.97), TXNL1 (1.82), RIN3 (1.95), CYBB (2.04), ZNF814 (1.72), KLF4 (1.28) etc. exceeds the value for $p = 0.1$. It indicate that this type of genes significant at the level of 90%. We have showed only the expression profile plots of genes of GDS-2643 data set (Fig. 2.).

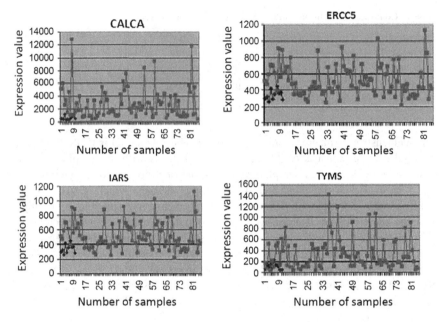

**Fig. 2** Expression profiles of some over-expressed genes (CALCA, ERCC, IARS) and under-express (TYMS) in normal (shown by *blue points*) and disease (shown by *red points*) samples of human leukemia expression data

## 3.1 Statistical Validation

Applying BMM-1 and BMM-2 we have found non-small leukemia and small leukemia pathway. In these two pathways a set of 472 genes is involved. This set of genes we have compared obtained by 3 methods. The result of BMM-1 and BMM-2, we have identified 297 and 307 number of genes are common in database information. We have said these genes are *true positive* (*TP*) genes. On the other hand we have found 104 and 100 number of genes that are in the set of 472 genes respectively which is obtained by BMM-1 and BMM-2 but not present in the pathway. These 104 and 100 number of gene are said *false positive* (*FP*) and the number of *false negative* (*FN*) gene is 100 and 103 for BMM-1 and BMM-2 respectively. Figure 3, compared to all other methods, it is comprehensible that both BMM-1 and BMM-2 have been efficient to identify more number of true positive genes but less number of false positive and false negative genes.

## 3.2 Biological Validation

The disease-mediating gene list and corresponding to an earmarked disease can be obtained in NCBI database [10] (http://ncbi.nlm.nih.gov/projects/geo/.). The list is composing in terms of relevancy of the gene. For leukemia, we have identified 349

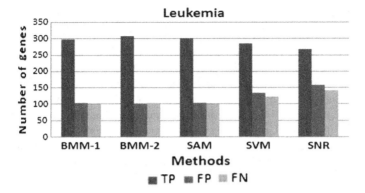

**Fig. 3** Comparison among the methods. Here *TP*, *FP*, *FN* indicate the number of true positive, false positive, false negative, respectively

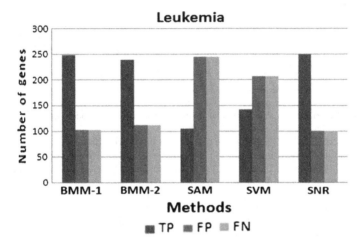

**Fig. 4** Comparison among the methods using NCBI database. Here *TP*, *FP*, *FN* indicate the number of true positive, false positive, false negative, respectively

numbers of genes each by using BMM-1 and BMM-2 respectively. This set of genes we have compared with 349 numbers of genes from NCBI and we can identified 247 and 238 numbers of genes for BMM-1 and BMM-2, respectively, which is common in both sets. We said that these genes are *true positive* (*TP*) genes. On the other hand (349 − 247) = 102 and (349 − 238) = 111 numbers of genes for BMM-1 and BMM-2, respectively, are not in the list which is obtained from NCBI. We said that these genes as *false positive (FP)*. Similarly (349 − 247) = 102 and (349 − 238) = 111 number of genes that are present in the NCBI list but not in the set of genes which is obtained by BMM-1 and BMM-2 respectively. In this reason, these genes are call *false negative (FN)*. Likewise, we have compared our results with other 3 methods, viz., SAM, SVM and SNR. Figure 4, show the corresponding results.

# 4 Conclusion

In this article, we have provided a procedure based on Boltzmann machine models for selection of genes that under or over expression may be normal or may be malignant. The procedure finds several groups of genes based on values of correlation. This is pursued by determining the most important group. Using BMM-1 and BMM-2 the genes of these groups are evaluated. The most important genes identified by the procedure have also been corroborated using their $p$-values. The performance of the procedure compared to few existing ones has reported. The results have been corroborated using biochemical pathway, $p$-value, $t$-test, sensitivity, and some existing result expression profile plots.

# References

1. Henry, D.: Latest News in Blood Cancer Research. CANCER Care, New York (2010)
2. Lim, G.: Overview of cancer in Malaysia. J. Clin. Oncol. **32**(1), 37–42 (2002) (Japanese)
3. Lv, J., Peng, Q., Chen, X., Sun, Z.: A multi-objective heuristic algorithm for gene expression microarray data classification. Expert Syst. Appl. **59**, 13–19 (2016)
4. Karakida, R., Okada, M., Amari, S.: Dynamical analysis of contrastive divergence learning: restricted Boltzmann machines with Gaussian visible units. Neural Netw. **79**, 78–87 (2016)
5. Yasuda, M., Horiguchi, T.: Triangular approximation for Ising model and its application to Boltzmann machine. Physica A **368**, 83–95 (2006)
6. Cawley, G.C., Talbot, N.L.C.: Gene selection in cancer classification using sparse logistic regression with Bayesian regularization. Bioinformatics **22**, 2348–2355 (2006)
7. Guyon, I., Weston, J., Barnhill, S.: Gene selection for cancer classification using support vector machines. Mach. Learn. **46**, 389–422 (2002)
8. Zhang, H.H., Ahn, J., Lin, X., Park, C.: Gene selection using support vector machines with non-convex penalty. Bioinformatics **22**, 88–95 (2006)
9. Goh, L., Song, Q., Kasabov, N.: A novel feature selection method to improve classification of gene expression data. In: Asia Pacific Bioinformatics Conference, Dunedin, New Zealand, vol. 29, pp. 161–166 (2004)
10. De, R.K., Ghosh, A.: Neuro-fuzzy methodology for selecting genes mediating lung cancer. In: 4th International Conference on Pattern Recognition and Machine Intelligence, pp. 388–393 (2011)
11. Liu, Y., So, R.M.C., Cui, Z.X.: Bluff body flow simulation using lattice Boltzmann equation with multiple relaxation time. Comput. Fluids **35**, 951–956 (2006)
12. National Center for Biotechnology Information. http://www.ncbi.nlm.nih.gov

# Survey of Classification Approaches for Glaucoma Diagnosis from Retinal Images

Niharika Thakur and Mamta Juneja

**Abstract** The eye is a vital and complex organ of vision that helps us in interaction with the outside world. It helps us to visualize the outside world by detecting light and converting it into impulses for sending them to the brain via the optic nerve. As it is sensitive in nature, so it is easily vulnerable to many diseases. Glaucoma is one of the second largest eye disease resulting in irreversible blindness, due to the damage of optic nerve. Ophthalmologists use retinal fundus images for assessment of this disease by manually outlining the optic cup and optic disc for analysis of the abnormality. The aim of this paper is to analyze different approaches used so far for classification of retinal images as abnormal or normal using feature extraction and classification.

**Keywords** Optic disc · Optic cup · Classification · Glaucoma

## 1 Introduction

The disease glaucoma came into existence in the initial years of seventeenth century, but its role as a source of blindness was accepted in nineteenth century. Since then its treatment has been initiated from twentieth century and is expected to be prevented by the end of twenty-first century. According to the ophthalmologists, the word glaucoma means blindness that appears in the initial years and looks like varnishing of the pupil. This word came from the ancient greek which means blue-green shade or cloud that describe the person with swollen cornea or rapid growth of cataract caused due to increased pressure inside the eye in cornea. During ancient times in early nineteenth century it was confused with cataract as both are

N. Thakur (✉) · M. Juneja
Computer Science and Engineering, University Institute of Engineering & Technology,
Panjab University, Chandigarh, India
e-mail: niharikathakur04@gmail.com

M. Juneja
e-mail: mamtajuneja@pu.ac.in

© Springer Nature Singapore Pte Ltd. 2018
R.K. Choudhary et al. (eds.), *Advanced Computing
and Communication Technologies*, Advances in Intelligent Systems
and Computing 562, https://doi.org/10.1007/978-981-10-4603-2_10

related with rise in the pressure inside the eye. In the European writings by Dr. Richard Bannister it was initially recognized as disease with four symptoms that were long duration of the eye disease, high tension of eye, presence of fixed size pupil and difficulty in perception of light [1]. It is a disease of eye in which devastation to the optic nerve of eye points to irreversible and progressive vision loss. The optic nerve carries visual information to brain from retina due to which we are able to locate the outside world. It is caused due to rise in the pressure of the eyes known as intraocular pressure. This pressure is increased when the amount of fluid produced in the eyes increases and as a result blockage occurs in the drainage that is the outflow channel. It is identified by damage to the optic nerve that starts with deteriorating vision and finally results in blindness. In many cases, the drainage of the eye becomes jammed due to which fluid cannot drain out through the eye and hence the pressure build up of fluid can inflict the optic nerve of eye and lead to dropping of vision. This usually affects both the eyes, but initially damage to one eye is more as compared to other eye [2]. According to the statistics of World Health Organization (WHO), it is the second most prevailing cause of irreversible blindness next to cataract. It has affected nearly 3 million people in the America out of which 1.5 million are unaware of it. 10% of people all over the world face problem of vision loss, even after the proper treatment. It cannot be cured completely once detected, but only its progression can be halted by medicines or surgeries to protect the eye from vision loss. Everyone from babies to senior citizens are at risk of glaucoma but older citizens are at higher risk of this disease. In the United States every 1 out of 10,000 infant babies are at risk of glaucoma. African Americans are mostly affected by this disease at a younger age. In case of glaucoma, there are no symptoms initially as there is no pain, also increase in the level of intraocular pressure does not affect the eye. But there may be the side vision loss which is initially not noticeable so the best way to protect eyes from this disease is to take immediate consultation of ophthalmologists when any kind of sight related problem is observed. People having diabetes and near sightedness are at high risk of this disease [3]. Prevention of this disease depends on the type of glaucoma that is detected by the experts and accordingly the suitable treatment given to the patient. Hence, it's accurate and timely detection can limit its progression not completely but to a certain extent. It is diagnosed by calculating the CDR ratio, ISNT Rule, DDLS (Disk damage likelihood scale), and GRI (Glaucoma risk index) which are achieved after extracting the optic cup and optic disc from retinal fundus images. Optic disc is the starting of optic nerve which is the location from where nerves of the retinal cells come close to each other. It is the entering point of blood vessels that passes blood to the retina. Whereas, optic cup is the central depression of variable size present on the optic disc. A pale disc is the indication of a disease condition which varies in color from orange to pink.

In this paper, various classification approaches used till date by different researchers for classification of retinal images as normal or abnormal have being discussed with their performance analysis. Each of them extracted features such as cup to disc ratio, gabor features, FFT coefficients, histogram features, principal components, and many more for classification.

# 2 Literature Survey

Some of the approaches used till date for classification of abnormality from retinal fundus images are as follows:

Classification is a decision theoretic approach to identify images by depicting one or more features of images and assigning it a label. In case of medical imaging, classification depicts the case as normal or abnormal based on analysis from some features extracted using classification approaches such as neural network, Support vector machine and many more [4].

Abramoff et al. [5] gave an automated approach of segmentation for optic disc using physiological plausible features. They initially cropped the image into $512 \times 512$ region of interest each with optic disc as center. Features of Gaussian filter bank with stereo disparity maps were then added for optimal use of color information in hue, saturation and brightness color spaces. Classification was carried out using "majority-win" of $k$-NN classifier after selection of features in sequential forward floating manner. Performance of the proposed approach gave 0.73, 0.81, and 0.86 correlation of cup-to-disc ratio and 0.93 correlation of proposed approach when compared with reference. But this approach was slow in speed and hence needs to be improved further for time computation [5].

Bock et al. [6] presented classification approach for diagnosis of glaucoma based on features extorted from retinal fundus images. Pre-processing of input image was carried out by homomorphic surface fitting followed by removal of blood vessels and normalization. Standard appearance based approach was then used to extract intensity values as feature vectors. Principal component analysis (PCA) and Linear discriminant analysis (LDA) was used as supervised and unsupervised approach for dimensionality reduction. For extraction of textural features Gabor filter banks were used on pre-processed image followed by PCA for feature reduction. For extraction of frequency components fast Fourier transform (FFT) was used followed by PCA for feature reduction. Finally, histogram was computed and Gaussian mixture model was applied for extraction of variance, mean, and weight that served as features to be classified. Then for the purpose of classification three classifiers namely naive Bayes, $k$-nearest neighbor and support vector machine were used with enhanced classification by applying feature selection and adaptive boosting to improve the results. Features were then combined to generate 2 stage classification probability score. Performance analysis showed that Support vector machine outperformed the naive Bayes and $k$-nearest neighbor with success rate of 86% [6].

Bock et al. [7] used two stage classification schemes that combine features to extract Glaucoma Risk index (GRI). They started with pre-processing in green channel of retinal fundus image to exclude features not related with disease glaucoma by illumination and reflectance transformation for illumination correction. For vessels impainting adaptive threshold followed by filtering with mask of canny was used. Centre of optic disc was achieved by intensity smoothing and threshold probing. Fourier coefficients, Pixel intensity values and B-spline coefficients were the features extracted for the classification of glaucoma. Finally, for classification two-stage Support vector machine was used with selected features as input to classifier. Glaucoma risk index (GRI), Cup-to-disk ratio (CDR), Glaucoma probability score (GPS), and Receiver operating characteristic (ROC) were evaluated for analyzing the performance of classifier. The proposed approach achieved accuracy of 80%, sensitivity of 73%, specificity of 85% and area under curve of 88% which was better than previously existing ones. But had a classification limitation in presence of low contrast images and increased disc size [7].

Xu et al. [8] proposed sliding window-based histogram feature detection approach for glaucoma diagnosis by cup detection. Localized disc was represented as non-rotated and arbitrary sized ellipse with center points and rectangular bounding box followed by generation of cup candidates using sampling. For feature representation, new region-based color concept was used in which values of green, blue, saturation and values were histogramed by the quantization with different number of bins such that every bin has equal number of pixels and channels. L1 normalized histogram of non cup regions and cup regions along with proportion of cup with respect to other pixels were taken as features. After histogram analysis one of the bins was categorized as pixels belonging to cup region. Then to select the desired features from large no. of features, statistical method of learning was applied to select features based on group sparsity constraint. Finally, for cup detection features were fed to kernelized SVM with LibSVM toolbox and non maximal suppression approach to reduce redundancy. Proposed approach outperformed the level set-based approach improving area under curve with value of 26% and CDR error of 0.09. But requires improvement in cup detection of large size [8].

Mookiah et al. [9] presented an improved classification approach for diagnosis of glaucoma by feeding discrete wavelet transform (DWT) and higher order spectra (HOS) as features to SVM Classifier with Glaucoma Risk Index (GRI). Initially for pre-processing, histogram equalization was applied to improve the contrast of image followed by radon transform to convert image projections into peaks. Different higher order spectra features such as phase entropy, bispectrum entropy1, entropy2, entropy3, average energy of wavelets were extracted and feed into the classifier. From the 54 discrete wavelet features, few of them were selected using average and mean values. Finally Support vector machine was used to classify the features as normal or abnormal and glaucoma risk index (GRI) was computed based on significance of features. On comparison of the recommended approach with existing approaches, it was found that performance of this approach was better than others with sensitivity of 93.33%, specificity of 96.67% and accuracy of 95%. But it can further be improved by varying the type of features [9].

Xu et al. [10] proposed a super pixel-based learning framework for diagnosis of glaucoma by optic cup detection. Simple linear iterative clustering was used to segment the retinal image into super pixels followed by bottom hat filtering to identify vessels mask and remove those pixels by superimposing it on input retinal fundus image. Position and color feature vectors such as normalized histogram were then selected for classification due to their relevance in this particular application. Apart from this, prior knowledge was initially used to train SVM classifier by LibLINEAR toolbox with weighted features. Contextual information was used to refine the labels and reduce classification error followed by ellipse fitting for final cup detection. On analysis of the result it was found that overlap ratio and CDR error improved as compared to previous approaches. But with increase in number of super pixel labels, over segmentation results into misclassification of the cup detection [10].

Cheng et al. [11] used a super pixel classification based approach for segmentation of optic cup and optic disc for glaucoma diagnosis. Centre surround statistics (CSS) was here computed to include feature that include difference between disc region and PPA region by generation of nine spatial scale dyadic gaussian pyramid which are low pass filtered channels. HIST and CSS were then combined to form a feature that was included for classification of optic disc. After that linear SVM was used to classify these features for glaucoma screening. Finally, binary matrix was created based on these decision values by using +1 for disc region and −1 for non disc region which average of 0. Morphological processing, elliptical Hough transform and active shape model was finally used to get optic disc boundary and reliability score was calculated for optic disc assessment. Cup used Distance between superpixels (D) along with HIST and CSS in addition to disc as features for classification [11].

Noronha et al. [12] gave another classification approach for glaucoma diagnosis using higher order spectrum cumulant features. For pre-processing of input retinal image interpolation and adaptive histogram equalization was applied to remove non uniform illumination. After this, radon transform was applied to convert image into projections along different angles followed by extraction of higher order spectrum cumulant features. Linear discriminant analysis (LDA) and principal component analysis (PCA) and was also applied to reduce the number of features followed by ranking of features using fisher's discrimination index. Finally, support vector machine with radial basis function (RBF) and naive bayesian classifier was used for classification of abnormal and normal cases with accuracy of 92.65% [12].

Rao et al. [13] proposed an optic disc and cup segmentation approach for diagnosis of glaucoma. They initially used Z-score normalization of input image for pre-processing followed by $k$-mean clustering for optic disc and optic cup segmentation. Separable filters were then applied across rows of one-dimensional discrete wavelet transform to convert it into two-dimensional discrete wavelet transform for feature extraction. Short and long windows were later applied across high and low frequencies for analysis of multiresolution. For selection of features to be used for training classifiers various averaging functions were used. Finally for classification, multilayer perceptron and artificial neural network were used to classify it as abnormal or normal case. On comparison of the classification results it was analyzed that classification accuracy of multilayer perceptron and backpropagation ANN came to be 97.6% which was better than naive Baye's classifier that came to be 89.6%. Hence, an improved detection system was formed which had capability to segment and classify both normal and abnormal disc and cup. But it still requires testing on other classifiers so as to improve the performance [13].

Tan et al. [14] used super pixel classification-based approach for localization of optic cup. This approach was helpful in reducing the effect of varying illumination between images and had capability of better boundary detection using multiple super pixel resolution. Optic cup was localized from optic disc image and was contrast normalized to extract vessels. Super pixels were then extracted and blood vessels were removed from it. For unique labeling of super pixels, classification models were integrated with sparse learning approach. They started with removal of blood vessels by applying multiple difference of closing with disk as structured element of varying radius. Histogram stretching and normalization was performed for pre-processing the image for accurate localization. Multi-scale super pixels were then extracted using simple linear iterative clustering (SLIC) approach, which combines pixels using $k$-means clustering. These features were then classified using support vector machines (SVM) classifier by training them with SLEP toolbox and $\alpha\beta$ base model. On testing the results, it was analyzed that using multiple model approach improved the performance. It was also found that the approach outperforms the previously existing approaches by improving the parameters such as area under curve of 0.95, average overlap error of 0.2, sensitivity of 85% and absolute CDR error of 0.081 but had a limitation of increased computation [14].

Sing et al. [15] presented an approach for classification of glaucomatous and non-glaucomatous retinal images. They took optic disc by localizing the center and segmenting it using bit plane analysis of 6th, 7th, and 8th bit plane after vessels impainting from morphological operations as input. Initially they extracted 18 features from segmented optic disc using first level wavelet analysis of discrete wavelet transform. Then from those extracted features, they only selected relevant ones by using principle component analysis and evolutionary attribute selection for further classification after $z$-score normalization. On these selected features, various classifiers such as decision tree, $k$-NN, K Star, Random forest, support vector machine (SVM), and artificial neural network (ANN) were used for classification to analyze which approach gives the best performance. Finally, classifier with best accuracy was considered for the purpose of classification. For the features extracted using evolutionary attribute selection without $z$-score normalization Random forest and ANN classifiers gave the highest accuracy of 94.75%. Whereas, for the features extracted using PCA with Z-score normalization $k$-NN and SVM gave highest accuracy of 94.75% [15].

Mahapatra et al. [16] proposed another segmentation approach for optic disc and optic cup using field of expert model. They started with pre-processing of optic disc using multiscale difference of closing algorithm for vessels impainting with disk structuring element. Elliptical Hough transform was then used to localize optic disc approximately. Square bounding box with size twice the optic disc diameter and center of optic disc was taken into consideration. Field of expert filters were used to extract features and were fed into Random forest classifier. Second order Markov random field was finally used for optimization using graph cut. Proposed approach achieves correlation of 0.85 with field of expert model which gives better results than other models [16].

Lotankar et al. [17] proposed another classification approach for glaucoma diagnosis using $k$-NN classifier. They evaluated features such as Horizontal to vertical CDR, vertical cup to disc ratio, cup to disc area ratio and rim to disc ratio. Finally, to classify they used Support vector machine, naive bayesian and $k$-NN classifier but results showed that classification accuracy was maximum in case of $k$-NN classifier i.e. 99% but it degraded its performance with increase in number of test images which can be considered as future work [17].

Table 1 compares various classification approaches used so far for glaucoma diagnosis with their performances.

**Table 1** Comparison of various classification approaches

| Authors | Features extracted | Classifiers used | Performance parameters |
|---|---|---|---|
| Abramoff et al. [5] | CDR, Gabor, and stereo features | $k$-NN classifier | Correlation of 0.93 |
| Bock et al. [6] | Intensity, FFT, and histogram features | SVM, Naive bayes, $k$-NN | Success rate of 86% |
| Bock et al. [7] | GRI, GPS, CDR, Fourier, and intensity coefficients | Two-stage SVM | Accuracy: 80% Area under curve: 88% |
| Xu et al. [8] | Histogram features | Kernalized SVM | Non overlap ratio: 26.8% Absolute area difference: 31.5% |
| Mookiah et al. [9] | HOS, DWT | SVM | Sensitivity: 93.33% Accuracy: 95% |
| Xu et al. [10] | Position and color features | SVM | Non-overlap ratio: 26.7% Relative area difference: 29% Absolute CDR error: 0.081 |
| Cheng et al. [11] | Histogram features, CSS, Distance measure, CDR | Linear SVM | Mean CDR error of 0.107 and 0.07 for normal and abnormal case Area under curve: 0.80 |
| Noronha et al. [12] | Higher order spectrum cumulant features | SVM, Naive bayes | Accuracy of 92.65% Sensitivity of 100% Specificity of 92% |
| Rao et al. [13] | DWT, CDR | Naive Bayes, MLP-BP/ANN | Accuracy of 89.96% for Naive Bayes classifier and 97.6% for MLP-BP/ANN classifier |
| Tan et al. [14] | Super pixels from SLIC, CDR | SVM | Area under curve: 0.95 Average overlap error: 0.2 Sensitivity: 85% Absolute CDR error: 0.081 |
| Sing et al. [15] | DWT, principal component | Random forest, ANN, $k$-NN, SVM | Accuracy: 94.75% for Random forest and ANN Accuracy: 94.75% for $k$-NN, and SVM |
| Mahapatra et al. [16] | Features extracted from field of expert model | Random forest classifier | Correlation of 0.85 |
| Lotankar et al. [17] | V-CDR, HV-CDR, CDAR and RDR | SVM, Naive bayesian and $k$-NN classifier | Sensitivity: 86% Specificity: 84% Accuracy: 99% for $k$-NN |

## 3 Conclusion

This paper presents different classification approaches used so far for detection of glaucoma. Based on the existing studies, it may be conclusively argued that SVM Classifier, the commonly used classifier for the detection of normal and abnormal cases has better specificity, sensitivity and accuracy as compared to other approaches. But still there remain residual challenges such as presence of retinal vessels and peripapillary atrophy that deteriorate the performance of the classification. It has been observed that researchers have considered number of features from time to time for the purpose of classification, but still there remains some features which can be taken into consideration for improved performance. Hence, an improved feature extraction approach is required followed by proper use of classifiers for the purpose of classification.

## References

1. Glaucoma: http://www.glaucoma.org.au/downloads/History.pdf
2. Bhowmik, D, et al.: Glaucoma—an eye disorder, its causes, risk factor, prevention and medication. Pharma J. **1**(1), 66–81 (2012)
3. Statistics of glaucoma: http://www.glaucoma.org/glaucoma/glaucoma-facts-and-stats.php
4. Classification in image processing: http://homepages.inf.ed.ac.uk/rbf/HIPR2/classify.htm
5. Abramoff, M.D., et al.: Automated segmentation of the optic disc from stereo color photographs using physiologically plausible features. Invest. Ophthalmol. Vis. Sci. **48**(4), 1665–1673 (2007)
6. Bock, R., et al.: Classifying glaucoma with image-based features from fundus photographs. In: Joint Pattern Recognition Symposium, pp. 355–364. Springer, Berlin (2007)
7. Bock, R., et al.: Glaucoma risk index: automated glaucoma detection from color fundus images. J. Med. Image Anal. **14**(3), 471–481 (2010)
8. Xu, Y., et al.: Sliding window and regression based cup detection in digital fundus images for glaucoma diagnosis. In: Springer Proceedings of International conference of Medical Image Computing and Computer-Assisted Intervention (MICCAI 2011), pp. 1–8 (2011)
9. Mookiah, M.R.K., et al.: Data mining technique for automated diagnosis of glaucoma using higher order spectra and wavelet energy features. J. Knowl. Based Syst. **33**, 73–82 (2012)
10. Xu, Y., et al.: Efficient optic cup detection from intra-image learning with retinal structure priors. In: Springer Proceedings of International conference on Medical Image Computing and Computer-Assisted Intervention (MICCAI 2012), pp. 58–65 (2012)
11. Cheng, J., et al.: Superpixel classification based optic disc and optic cup segmentation for glaucoma screening. IEEE Trans. Med. Imaging **32**(6), 1019–1032 (2013)
12. Noronha, K.P., et al.: Automated classification of glaucoma stages using higher order cumulant features. J. Biomed. Signal Process. Control **10**, 174–183 (2014)
13. Rao, P.V., et al.: A novel approach for design and analysis of diabetic retinopathy glaucoma detection using cup to disk ration and ANN. Procedia Mater. Sci. **10**, 446–454 (2015)
14. Tan, N.M., et al.: Robust multi-scale superpixel classification for optic cup localization. Comput. Med. Imaging Graphics **40**, 182–193 (2015)
15. Sing, A., et al.: Image processing based automatic diagnosis of glaucoma using wavelet features of segmented optic disc from fundus image. Comput. Methods Progr. Biomed. **124**, 108–120 (2015)

16. Mahapatra, D., Buhmann, J.M.: A field of experts model for optic cup and disc segmentation from retinal fundus images. In: IEEE Proceedings of 12th International Symposium on Biomedical Imaging (ISBI 2015), pp. 218–221 (2015)
17. Lotankar, M., Noronha, K., Koti, J.: Detection of optic disc and cup from color retinal images for automated diagnosis of glaucoma. In: Proceedings of IEEE UP Section Conference on Electrical Computer and Electronics (UPCON 2015), pp. 1–6 (2015)

# Discovering Optimal Patterns for Forensic Pattern Warehouse

Vishakha Agarwal, Akhilesh Tiwari, R.K. Gupta
and Uday Pratap Singh

**Abstract** As the need of investigative information is increasing at an exponential rate, extraction of relevant patterns out of huge amount of forensic data becomes more complex. Forensic pattern mining is a technique that deals with mining of the forensic patterns from forensic pattern warehouse in support of forensic investigation and analysis of the causes of occurrence of an event. But, sometimes those patterns do not provide certain analytical results and also may contain some noisy information with them. An approach through which optimal patterns or reliable patterns are extracted from forensic pattern warehouse which strengthen the decisions-making process during investigations has been proposed in the paper.

**Keywords** Data mining · Forensic investigation · Genetic algorithms · Pattern warehousing · Optimal forensic patterns

## 1 Introduction

Internet is nowadays preferred as the most significant option for facilitating common man with enormous information. In order to handle this emerging data, several types of repositories are being introduced like databases, data warehouse [1], and

V. Agarwal (✉) · A. Tiwari · R.K. Gupta
Department of CSE & IT, Madhav Institute of Technology and Science,
Gwalior, India
e-mail: agarwal.vishakhacse@gmail.com

A. Tiwari
e-mail: atiwari.mits@gmail.com

R.K. Gupta
e-mail: iiitmrkg@gmail.com

U.P. Singh
Department of Applied Mathematics, Madhav Institute of Technology
and Science, Gwalior, India
e-mail: usinghiitg@gmail.com

© Springer Nature Singapore Pte Ltd. 2018                                      101
R.K. Choudhary et al. (eds.), *Advanced Computing
and Communication Technologies*, Advances in Intelligent Systems
and Computing 562, https://doi.org/10.1007/978-981-10-4603-2_11

pattern warehouse. Each of these repositories contains information at some consolidated level. Pattern Warehouse is a kind of repository which stores the data in the form of patterns, which is a knowledge representative.

## 1.1  Background and Gaps in the Current Scenario

Bartolini et al. in [2] suggested the concept of patterns and pattern warehousing and also drew a conceptual architecture for pattern base management system. Later on, other researchers [3–7] also contributed in the same domain at architectural, structural, and query processing level.

Recently Tiwari et al. [8] coupled the concept of pattern warehousing with forensic domain and introduced a new repository called as forensic pattern warehouse for storing forensic patterns. The focus of the author was to design a system which can store the patterns extracted out of the forensic databases and to develop a technique which performs forensic examination and analysis upon those patterns and generate knowledge which directly helps in further investigative decision-making process. This approach also addresses following issues which may lead to derive spurious results.

- The architecture proposed in the literature for forensic pattern warehousing has some important components missing, without which implementation would not be possible.
- The illustrative aspect from their logic model is also absent.
- The feature of reliability is also absent, i.e., among the extracted patterns there may be some false forensic patterns.

So, after analyzing all the above issues, author in this paper, has taken into consideration the pattern warehouse of forensic domain, upon which pattern mining could be performed for finding optimal or reliable patterns using some standard heuristic approaches. Furthermore, author took dataset containing circumstances of various accidents that are occurring so frequently in a city. Now, the objective of the paper is to provide investigative help to the analyst through optimal pattern mining by discovering the patterns which contain information regarding the most appropriate causes of these accidents so that further remedial actions could be taken by the investigators. Figure 1 gives an overview of the process proposed for finding optimal patterns from forensic data warehouse.

So, in support of this, author in this paper, proposes a new conceptual architecture for finding optimal patterns for forensic pattern warehouse and also developed an algorithm which filters optimal patterns out of forensic pattern warehouse.

**Fig. 1** Pyramid depicting the
process of finding optimal
patterns from forensic data
warehouse

## 2 Proposed Architecture for Optimal Forensic Pattern Warehousing

Analyzing the above problem, an architecture has been proposed in Fig. 2 for the forensic pattern warehousing, which eventually finds the optimal patterns out of the forensic pattern warehouse. The various layers of the proposed architecture are responsible for the followings tasks as follows:

1. *Physical Layer*—This is the lowest basic layer which incorporates the data warehouses of the forensic domain, i.e., forensic data warehouses and the forensic data marts which are acting as the source of forensic domain data in this layer.
2. *Pattern Generation Layer*—This layer includes the pattern generation engine which incorporates all the techniques, tools, and approaches for finding the patterns out of these forensic data repositories. These approaches differ according to the pattern type needed by the forensic pattern analyst.
3. *Pattern Warehousing Layer*—Now, in this third layer all the forensic patterns extracted out from the forensic data warehouse by the above engine are stored in a nonvolatile manner in a repository called as forensic pattern warehouse. This layer also contains forensic pattern marts which are responsible for holding the forensic patterns either of specific type or specific department for an organization.
4. *Optimization Layer*—In this layer, author has incorporated a new engine specifically for refining the forensic patterns within the forensic pattern warehouse. In this layer, this engine integrates genetic-based optimization approach.
5. *Application Layer*—This is the topmost layer which provides the interface to the forensic pattern analyst for isolating the analysis results from the optimal forensic patterns. The result of the previous layer, i.e., filtered patterns is visible in this layer through which forensic reports and analytical results are generated.

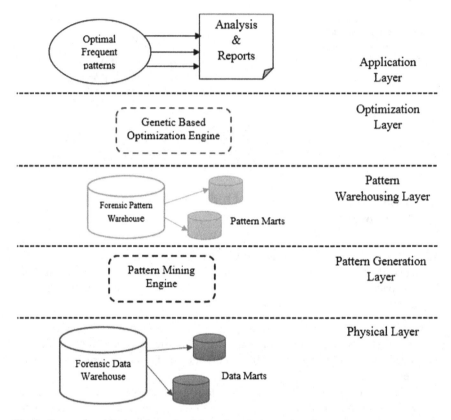

**Fig. 2** Proposed architecture for optimal forensic pattern warehousing

## 2.1 Illustrative Aspect of Optimal Forensic Pattern Warehousing

To illustrate this architecture author has taken a synthetic dataset (Table 2) consisting of the circumstances under which different accidents occurred. This dataset is an instance of a forensic data warehouse which contains this information regarding various cities. Slicing is performed and data related to the various accidents occurred in a city is extracted, based on which the forensic pattern analyst will analyze the patterns depicting the major cause of these accidents. Six attributes have been selected for depicting the circumstances of the accidents occurred in a city (Table 1).

Dataset containing the information of six accidents.

Now, frequent pattern mining algorithm is applied over this forensic dataset to mine the forensic patterns out of this forensic dataset. These are shown in Table 3.

Table 3 shows the instance of the forensic pattern warehouse. These patterns extracted by the pattern mining engine are stored in a volatile manner in forensic

**Table 1** Various attributes and their values that are considered in dataset

| Attributes | Values |
|---|---|
| A. Location | A1 Roundabout<br>A2 Tunnel<br>A3 Road works<br>A4 Bridge |
| B. Weather conditions | B1 Rain<br>B2 Fog<br>B3 Stormy<br>B4 Normal Weather |
| C. Light conditions | C1 Night<br>C2 Twilight<br>C3 Daylight |
| D. Driver condition | D1 Ill<br>D2 Sedated<br>D3 Drunk<br>D4 Normal |
| E. Driver's age | E1 18–29<br>E2 30–45<br>E3 46–60<br>E4 above 60 |
| F. Obstacle type | F1 Animal<br>F2 Street Car<br>F3 Against Crash Barrier<br>F4 Speed Ramp |

**Table 2** Sample forensic dataset

| Accident Id | Values |
|---|---|
| AC1 | A1, B2, C1, D3, E2, F1 |
| AC2 | A1, B2, C1, D3, E3, F3 |
| AC3 | A1, B2, C1, D1, E3, F4 |
| AC4 | A1, B2, C3, D3, E1, F2 |
| AC5 | A1, B1, C1, D3, E4, F4 |
| AC6 | A2, B2, C1, D3, E2, F2 |

pattern warehouse upon which further filtering or optimization process will be performed.

Now, the genetic-based optimization engine starts functioning, i.e., filtering the patterns based upon the optimization algorithm which author has discussed below.

---

**Algorithm: GA-based optimization approach**

**Input**: Frequent forensic pattern set, crossover rate, mutation rate.
**Output**: Optimal forensic patterns.
1. Initialize from the random population Pac of frequent forensic patterns.
2. **While** (whole population converges) **do**
   (i) Select two parents from $P_{ac}$;
   (ii) Perform crossover over selected pair of parents;

---

(continued)

(continued)

(iii) Perform mutation and get the new population;
(iv) Insert the offspring to $P'_{ac}$.
$P_{ac} \leftarrow P'_{ac}$
(v) Test the fitness of each chromosomes in the new population;
**End while**
3. Return with optimal forensic patterns.

Now, after going through this algorithm the whole frequent forensic pattern set will be filtered and the optimal forensic patterns are mined out (Table 4) which provides much better results to the forensic pattern analyst.

As per the genetic algorithm, the convergence condition is user-specified which is number of frequent forensic patterns, i.e., until all the patterns are selected and genetically treated by the algorithm and it can be represented as

$$\text{No. of iterations} = |P_{ac}|/2$$

**Table 3** Frequent forensic patterns

| Pattern ID | Pattern | Support_value |
|---|---|---|
| P1 | A1 | 5 |
| P2 | B2 | 5 |
| P3 | C1 | 5 |
| P4 | D3 | 5 |
| P5 | A1, B2 | 4 |
| P6 | A1, C1 | 4 |
| P7 | A1, D3 | 4 |
| P8 | B2, D3 | 4 |
| P9 | B2, C1 | 4 |
| P10 | C1, D3 | 4 |
| P11 | A1, B2, C1 | 3 |
| P12 | A1, B2, D3 | 3 |
| P13 | A1, C2, D3 | 3 |
| P14 | B2, C1, D3 | 3 |

**Table 4** Optimal forensic patterns

| Pattern ID | Pattern |
|---|---|
| P1 | A1 |
| P2 | B2 |
| P5 | A1, B2 |
| P10 | C1, D3 |
| P11 | A1, B2, C1 |
| P12 | A1, B2, D3 |
| P13 | A1, C1, D3 |
| P14 | B2, C1, D3 |

# 3   Result Analysis

After executing the above algorithm over the forensic dataset, it has been observed that before applying the optimization algorithm, the number of frequent patterns that were stored in pattern warehouse are quite in large number but application of the genetic-based algorithm narrowed down the number of patterns, i.e., optimal patterns to a smaller number. This difference is best explained by the following graph (Fig. 3) which is representing the difference in number of optimal forensic patterns and frequent forensic patterns when optimization algorithm is applied over the forensic pattern warehouse.

Also, it almost eliminates all those patterns in which there is only one attribute which is acting as a cause for the accident and also the strength of that attribute is weak because in this investigation author supposes that there has to be a number of attributes which are creating the circumstances for the accident. For instance, consider the pattern $P12$ $\{A1, B2, D3\}$, this pattern shows that circumstance which lead to the accident is foggy weather in which a drunk driver was turning vehicle at the roundabout. Now, Table 5 points out the major differences that critically analyze the proposed approach with past approaches.

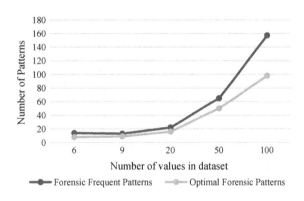

**Fig. 3** Graph representing the difference in number of frequent and optimal forensic patterns extracted out of a dataset

**Table 5** Key differences between proposed approach and past approaches

| Features | Proposed approach | Past approaches |
|---|---|---|
| Processing engine | Pattern mining engine and genetic based optimization engine | Data mining engine/knowledge discovery engine |
| Outcome | Optimal forensic patterns | Forensic patterns |
| Concept of pattern mart | Specified | Unspecified |
| Reliability of patterns | More reliable and optimal patterns | Presence of false patterns |
| Techniques for generating patterns | Soft computing integrated data mining approach for extracting patterns | Traditional data mining approaches for finding patterns |

# 4 Conclusion

Pattern warehousing is a complex activity, concerning any technique that can effectively generate, store, and manipulate patterns and also derive valuable information out of it. The author here presented application aspect of pattern warehousing and also in addition to that proposed a technique through which optimal patterns are mined out of the pattern warehouse. The inclusion of genetic algorithm provided strength to the current approach and also it does not add complexity to the implementation aspect of the algorithm. The inclusion of genetic algorithm with pattern mining is a quite novel work in this context. In future, incorporation of other heuristic approaches can be made in order to find more interesting patterns for forensic investigation and analysis.

# References

1. Han, J., Kamber, M.: Data Mining: Concepts and Techniques. Morgan Kaufmann, San Francisco (2006)
2. Bartolini, I., Bertino, E., Catania, B., Ciaccia, P., Golfarelli, M., Patella, M., Rizzi, S.: Patterns for next-generation database systems: preliminary results of the PANDA Project. In: 11th Proceedings of Italian Symposium on Advanced Database Systems, pp. 1–8. Cetraro (CS), Italy (2003)
3. Rizzi, S.: UML-based conceptual modeling of pattern-bases. In: Proceedings of the International Workshop on Pattern Representation and Management, pp. 1–11. Hellas (2005)
4. Terrovitis, M., Vassiliadis P., Skiadopoulos, S.: Modeling and language support for the management of pattern-bases. Data Knowl. Eng. 368–397. Elsevier (2007)
5. Evangelos, E., Kotsifakos, E.: Pattern-miner: integrated management and mining over data mining models. In: 14th Proceedings of ACM SIGKDD International Conference on Knowledge Discovery and Data Mining, pp. 1081–1084 (2008)
6. Tiwari, V., Thakur, R.S.: P2ms: a phase-wise pattern management system for pattern warehouse. Int. J. Data Mining Model. Manag. 5, 1–10, Inderscience (2014)
7. Tiwari, V., Thakur, R.S.: Contextual snowflake modeling for pattern warehouse logical design. In: Sadhana–Academy Proceedings in Engineering Science, vol. 39. Springer, Berlin (2014)
8. Tiwari, V., Thakur, R.S.: Improving knowledge availability of forensic intelligence through forensic pattern warehouse (FPW). In: Khosrow-Pour, M. (eds.) Encyclopedia of information science and technology, pp. 1326–1335. IGI Global, Hershey (2015)

# Part II
# Advanced Computing: Distributed Computing

# Comparative Study of Scheduling Algorithms in Heterogeneous Distributed Computing Systems

**Mamta Padole and Ankit Shah**

**Abstract** It is the need of an era to store and process big data and its applications. To process these applications, it is inevitable to use heterogeneous distributed computing systems (HeDCS). The heterogeneous distributed systems facilitate scalability, an essential characteristic for big data processing. However, to implement the scalable model, it is essential to handle performance, efficiency, optimal resource utilization and several other key constraints. Scheduling algorithms play a vital role in achieving better performance and high throughput in heterogeneous distributed computing systems. Hence, selection of a proper scheduling algorithm, for the specific application, becomes a critical task. Selection of an appropriate scheduling algorithm in heterogeneous distributed computing systems require the consideration of various parameters like scheduling type, multi-core processors, and heterogeneity. The paper discusses broadly the hierarchical classification of scheduling algorithms implemented in heterogeneous distributed computing systems and presents a comparative study of these algorithms, thus providing an insight into the significance of various parameters that play a role in the selection of a scheduling algorithm.

**Keywords** Scheduling algorithms · Heterogeneous distributed computing systems (HeDCS) · Scheduling in heterogeneous distributed systems · Dynamic scheduling · Big data processing

M. Padole (✉)
Department of Computer Science & Engineering,
The M.S. University of Baroda, Vadodara, India
e-mail: mpadole29@rediffmail.com

A. Shah
Department of Information Technology, SVBIT, Gandhinagar, India
e-mail: shah_ankit101@yahoo.co.in

© Springer Nature Singapore Pte Ltd. 2018
R.K. Choudhary et al. (eds.), *Advanced Computing
and Communication Technologies*, Advances in Intelligent Systems
and Computing 562, https://doi.org/10.1007/978-981-10-4603-2_12

# 1    Introduction

A distributed computing system (DCS) is a collection of processors that are connected to each other using network, communicate to each other by message passing
and used for the execution of resource-intensive applications [1]. Traditionally DCS
involves computing or processing using spatially distributed systems [2] implemented over the homogeneous set of computers, each having similar hardware
architecture and operating system, also referred as homogeneous distributed computing system (HDCS).

Contemporary distributed systems are implemented on machines having different
processors, memory, and an operating system, each connected via a high-speed
network. Such distributed computing system with different type of hardware architecture and the operating system is known as heterogeneous distributed computing
system (HeDCS), also referred as heterogeneous distributed system (HeDS).

Scheduling is a mechanism of allocating processors for execution of a task.
Tasks or resources are arranged in a manner so as to complete the work in minimum
time. Scheduling algorithms in homogeneous computing systems whether parallel
or distributed, usually run on uniform and dedicated set of processors [3].

The paper is structured as follows: Key Parameters of scheduling have been
discussed in Sect. 2. The Hierarchical classification of scheduling algorithms has
been discussed in Sect. 3. In Sect. 4, Comparison of Scheduling algorithms is
discussed, followed by the Conclusion in Sect. 5.

# 2    Scheduling Parameter

Scheduling in DCS is primarily concerned with two aspects such as optimizing
completion time [4] of an application and optimizing the resource utilization. In the
context of an application, the main parameter is to reduce the total cost of executing
a particular application whereas, optimal utilization and performance of the
resource is the prime concern of the resource provider. The two main factors in
defining the best performance of a scheduling algorithm in HeDCS are
application-specific and system-specific [5]. Thus, objective functions of scheduling
algorithms can be categorized into two broad classifications: Application-Specific
and System-Specific. Figure 1 displays the objective functions of scheduling
algorithms covered in this paper.

## 2.1    Application-Specific

Various scheduling parameters need to be considered while implying
application-specific scheduling. Application-specific scheduling explicitly addresses

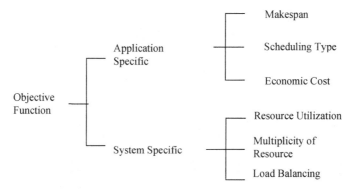

**Fig. 1** Objective function for scheduling algorithms

heterogeneity and conflict in distributed environments. Watchful scheduling of application components is essential to accomplish its performance objectives. Scheduling decisions are determined based on parameters like application performance, computational requirements, task inter-dependency, processing load and the availability of resources. Depending upon an application, parameters to be considered, may vary, to optimize the performance of a specific application.

## 2.2 System-Specific

In system-specific objectives, the main aim is resource utilization, particularly that of processors and memory. The variance in performance of the resources has a direct influence on the performance of the submitted application and must be deliberated during scheduling. Resource utilization, i.e., the percentage of time a resource is busy or available is of vital importance. Overutilization of a scarce resource means nonavailability of resource when the application needs it. This may increase the application waiting time, thus resulting in higher completion time. Other resource-specific objectives are load balancing, fixed number of processors, unbounded number of processors, etc.

The factors considered in this research paper are scheduling type, multi-core processors, heterogeneity, degree of multiprogramming, makespan, load balancing, multiplicity of resources, impact of bounded number of processors (BNP) and unbounded number of heterogeneous processors, optimized resource time, etc.

## 3  Hierarchical Classification

It is proven that the complexity of a scheduling problem is NP-Complete [6]. The scheduling problems are difficult to deal with, as different parameters have different impact during scheduling of tasks. This section discusses the hierarchy of

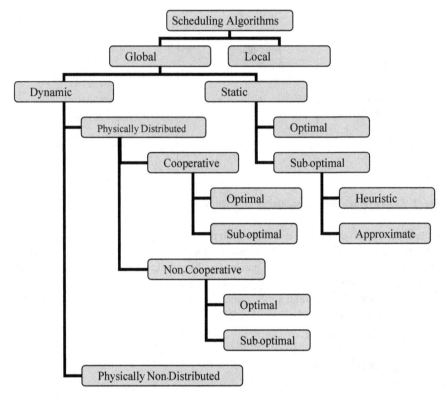

**Fig. 2** Hierarchical classification of scheduling algorithms

scheduling algorithms (Fig. 2) in distributed computing, which will aid in discussing issues related to the scheduling algorithms. Scheduling algorithms in distributed computing may be defined as a subset of this taxonomy. The subcategories of scheduling techniques can be identified as what follows [7].

Local versus Global—Local scheduling is implemented in uniprocessor environment whereas global scheduling in multiprocessor environment. In local scheduling, all the processes that need to be executed are assigned to a single CPU, on a single computer; whereas a global scheduling discipline executes the task by allocating or scheduling on multiprocessing system or in distributed system. As distributed scheduling falls into the global scheduling, local scheduling is not considered for discussion.

Static versus Dynamic—Further subcategorization of global scheduling is static or dynamic scheduling. It specifies the time when scheduling is decided. Static scheduling will be implemented when all information about tasks and resources are known before execution begins, whereas, in dynamic scheduling, the tasks are scheduled, as and when they arrive at runtime. Dynamic scheduling is advantageous when execution time is not known at the outset, the number of tasks, number of iterations in the task or the arrivals of tasks are random. It supports nondeterminism. Both static and dynamic scheduling are widely adopted in distributed computing.

Optimal versus Suboptimal—When all the information regarding the resource requirement of an application is known apriori, an optimal scheduling is a better option. Optimal scheduling is based on some criterion function, like makespan, resource requirement, resource sharing, and concurrency control. But due to the fact that scheduling algorithms are NP-complete [6], it is difficult to identify the resource requirements in advance. This may cause computational infeasibility. Suboptimal algorithms can be further categorized as heuristic and approximate.

Approximate versus Heuristic—An approximate scheduling algorithms are usually meant to give an approximate solution, with some kind of performance guarantee based on available parameters like minimize makespan, optimize performance, processor utilization, etc. The heuristic approach represents the category of static and dynamic algorithms. These algorithms help in accurately presuming the requirement of process and system load. By this method, the scheduling cannot be optimal but gives the most reasonable amount of time to perform the task.

Distributed versus Non-distributed—Under dynamic scheduling comes physically distributed or physically non-distributed schedulers.

It involves whether the task execution should be done on a single processor (physically non-distributed or centralized) or whether the task should be physically distributed (decentralized or spatially distributed) among the remotely placed processors. The centralized policy is easy to implement but it suffers from the lack of scalability, single point failure, and performance bottleneck. On the other hand, decentralized scheduling overcomes the problem of centralized system and also be implemented for scalability, load balancing, resource utilization and other advantages.

Cooperative versus Non-cooperative—In the realm of global, dynamic, physically distributed scheduling algorithms; the processors available for execution may be working cooperatively or independently. In non-cooperative case, individual processors work autonomously and take their decisions independently without concern of the rest of the system. In contrast, in cooperative scheduling, the processors involved are working for a single goal, in a cooperative manner. In cooperative scheduling each processor works in coordination with other processors, to take various scheduling decision and thus, to optimize throughput of the system.

In addition to the hierarchy discussed in the paper, there are so many other characteristics that distributed systems may have. All the branches that have been discussed here may not fit uniquely, but they are vital to describe the behavior of the scheduler. The classification tree is further extensible and is non-exhaustive.

# 4 Comparative Study

The study emphasizes on two aspects, one to find the objective behind using specific scheduling technique, and second to discuss the merits and possible enhancements to each technique. In Table 1, various algorithms have been listed.

**Table 1** Comparison of various scheduling algorithms

| Sr. No | Algorithm | Key parameter | Scheduling type | Nature of task | Environment | Objective | Comments | Algorithm compared with |
|---|---|---|---|---|---|---|---|---|
| 1 | Monte Carlo based DAG[a] scheduling approach [8] | Makespan, heterogeneity, bounded number of processors | Global, static, sub-optimal, heuristic | Flow of work | Dynamic DAG simulator | To minimize the makepan and optimize overall performance | Avoid the complex computation with random variable and applicable to any random distribution.DAG HEFT based which is less effective | – |
| 2 | SD[a]-based algorithm for task scheduling —SDBATS [9] | Makespan, multi-core processors | Global, static | Periodic task | Not mentioned | Effective schedule length and speed up | Suitable for some real-world application like Gaussian elimination and Fourier transformation | – |
| 3 | HEFT[a], MCP[a], ETF[a], HLEFT[a], DLS[a] [10] | Makespan, degree of multiprogramming | Global, static, sub-optimal, heuristic | Simultaneous tasks | Bounded number of homogeneous processors | To minimize make span | Standard algorithms for static task scheduling | – |
| 4 | Clustering for minimizing the worst schedule length—CMWSL [11] | Schedule length | Global, static | Task clustering | Not mentioned | To minimize Worst Schedule Length (WSL) | Suitable to data intensive application and heterogeneous system support | Clustering based task scheduling algorithm |
| 5 | Clustering based HEFT with duplication—CBHD [12] | Makespan, load balancing, optimize sleek time | Global, dynamic, distributed | Grouped task (cluster) | Distributed algorithm simulator | To minimize the execution time & provide load balancing. | Minimize makespan, maximize processor utilization by load balancing | HEFT, triplet, cluster |

(continued)

**Table 1** (continued)

| Sr. No | Algorithm | Key parameter | Scheduling type | Nature of task | Environment | Objective | Comments | Algorithm compared with |
|---|---|---|---|---|---|---|---|---|
| 6 | Multi queue balancing—MQB [4] | Makespan, load balancing | Global, dynamic, distributed | Parallel tree workload based | Discrete-time simulator | Reduce exec; time by max. resource utilization | Reduces the processor idle time and full use of resources | Compared with various types of load |
| 7 | Heterogeneous scheduling algorithm with improved task priority—HSIP [13] | Schedule length ratio, efficiency | Global, static | Simultaneous task | Heterogeneous simulation | Improve task priority strategy for optimized makespan | Algorithms well suitable for heterogeneous environment and useful for real world problems | PEFT, SDBATS, HEFT, CPOP[a] |
| 8 | Performance effective genetic algorithm—PEGA [14] | Optimal mapping, sequence of execution | Global, static, optimal | Random task | Heterogeneous simulation | To minimize finish time along with increase throughput | Algorithm outperforms against traditional scheduling methods. Not suitable for large data sets | RR[a], SJF[a], FCFS[a] |
| 9 | Hybrid genetic algorithm—HGA [15] | Makespan, Load balancing | Global, static, optimal | Regular/random task | Heterogeneous simulation | Optimization of makespan and utilize maximum resources | Follows Montage, CyberShake benchmark for validate performance | MCP, HEFT, PEGA, MPQGA, HS CGS[a] |
| 10 | Distributed QoS aware scheduler [16] | Latency, availability, system utilization | Global, dynamic, adaptive scheduler | Distributed stream processing | Peersim simulator | Improve performance by adaptive scheduler for distributed environment | Storm based scheduler for overall system performance and quality of services | cRR[a], cOpt[a] |

(continued)

**Table 1** (continued)

| Sr. No | Algorithm | Key parameter | Scheduling type | Nature of task | Environment | Objective | Comments | Algorithm compared with |
|---|---|---|---|---|---|---|---|---|
| 11 | Predict earliest finish time—PEFT [17] | Efficiency, makespan, better schedule length ratio | Global, static | Workflow | Not mentioned | Forecast the cost table and schedule the task accordingly to achieve low complexity | Suitable when all tasks are known along with their complexity and dependency | Look-ahead, HEFT,HCPT[a], PETS[a], HPS[a] |
| 12 | BDSC hierarchical BDSC—HBDS [18] | Resource management, speedup, bounded number | Global, static, sub-optimal, heuristic | Parallel task execution | PIPS platform | Effective resource management and speedups on shared and distributed environment | Suitable for signal processing, image processing application. For BNP system is robust | HLFET, ISH, MCP, HEFT |
| 13 | Stochastic dynamic level scheduling—(SDLS) [19] | Makespan, speedup, and nakespan | Global, dynamic, distributed | Precedence constrained tasks | Simulation cluster | To achieve better performance by dynamic scheduling | Effective for random task arrival time. This will be more effective by parallelism | HEFT, Rob-HEFT, andSHEFT |
| 14 | P-HEFT [20] | Makespan, heterogeneity, parallel task | Global, dynamic | Simultaneous tasks | Not mentioned | To minimize the completion time by parallel execution. | Suitable for image processing application. Best for multiple-mixed jobs. | Heterogeneous parallel task scheduler (HPTS) |
| 15 | Online scheduling of dynamic task graphs [21] | Number of processors, memory | Global, dynamic | Real time tasks | Not mentioned | To provide runtime scheduling | Interprocessor communications of fixed number of homogeneous | Multiprocessor scheduling algorithm |

(continued)

**Table 1** (continued)

| Sr. No | Algorithm | Key parameter | Scheduling type | Nature of task | Environment | Objective | Comments | Algorithm compared with |
|---|---|---|---|---|---|---|---|---|
| 16 | Self adaptive reduce scheduling—SARS [22] | Reduce completion time | Global, dynamic, distributed | Batch processing | Hadoop map-reduce cluster | To minimize average completion time and response time | Suitable to big data application processing | FIFO, fair, capacity scheduler |

[a]DAG Directed Acyclic Graph, *SD* Standard Deviation, *HEFT* Heterogeneous Earliest Finish Time, *MCP* Modified Critical Path, *HLEFT* Highest Level First Estimate Time, *DLS* Dynamic Level Scheduling, *GA* Genetic Algorithm, *ETF* Earliest Time First, *ISH* Insertion Scheduling Heuristic, *HCPT* Heterogeneous Critical Parent Trees, *PETS* Performance Effective Task Scheduling, *HPS* High Performance Task Scheduling. *SHCP* Scheduling with Heterogeneity using Critical Path, *HHS* Hybrid Heuristic Scheduling, *RR* Round Robin, *SJF* Shortest Job First, *FCFS* First Come First Serve, *BDSC* Bounded Dominant Sequence Clustering, *MPQGA* Multiple Priority Queues Genetic Algorithm, *HSCGS* Hybrid Successor Concerned Heuristic-Genetic Scheduling, *cRR* Centralized Round-Robin, *cOpt* Centralized Optimal scheduler, *FIFO* First In First Out

**Zheng et al.** [8] proposed Monte Carlo based Directed Acyclic Graph scheduling approach with the objective to minimize the makespan for BNP. This approach works well for any random distribution under heterogeneous environment. This approach gives competitive advantage compared to other static-heuristic techniques.

**Ehsan et al.** [9] proposed standard deviation-based algorithm for task scheduling (SDBATS) to reduce schedule length and speedup the scheduling by assigning task priority.

**Kwok et al.** [10] proposed to optimize the makespan by considering a wide range of techniques, genetic algorithm, randomization branch-and bound and graph theory. Authors have proposed many useful static, heuristic algorithms (e.g., HEFT, MCP, ETF, and DLS) but that will not be effective in today's era of big data.

**Kanemitsu et al.** [11] proposed clustering-based task scheduling algorithm that minimizes the schedule length for heterogeneous processors. It is apt for data-intensive application and has proven to be better than other list-based and clustering-based task scheduling algorithms.

**Abdelkader et al.** [12] proposed dynamic task scheduling algorithm for heterogeneous systems called Clustering-Based HEFT with Duplication (CBHD). This algorithm targets the three important parameters for getting better performance, minimize the makespan, load balancing and optimize the sleek time.

**Yuxiong et al.** [4] proposed Multi-queue Balancing (MQB) algorithm that minimizes the makespan and maximize the heterogeneous resource utilization. MQB has multiple queues for online scheduling to achieve better utilization and minimizing completion time.

**Wang et al.** [13] proposed heterogeneous scheduling algorithm with improved task priority (HSIP) for improvising schedule length ratio and task priority. This algorithm performs two-step process first, identifies the task priority and second finds the best processor to execute the tasks.

**Ahmad et al.** [14] proposed performance effective genetic algorithm (PEGA) which operates through large search space and finds the best solution using reproduction concept. Reproduction uses two operators namely crossover and mutation to select a random task and performs fitness function on it to select the best task to execute on the heterogeneous parallel multiprocessor system.

**Ahmad et al.** [15] proposed hybrid genetic algorithm (HGA) is a hybrid combination of HEFT heuristic and PEGA genetic algorithm. It provides optimize makespan and load balancing over heterogeneous systems.

**Valeria, et al.** [16] proposed distributed QoS-aware scheduling with self-adaptive capability in storm. By using this concept authors tried to overcome the limitation of high latency, less availability, and poor system utilization in distributed data stream processing (DSP).

**Hamid et al.** [17] proposed Predict Early Finish Time (PEFT) to speedup and optimize the makespan. It has two phases: a task prioritizing and a processor selection which identifies the task priority and allocates it to the best processor respectively.

**Khaldi et al.** [18] proposed static-heuristic scheduler called bounded dominant sequence clustering (BDSC) is an extension of DSC limiting the memory constraints and the bounded number of processors. It is suitable for signal processing and image processing kind of application.

**Kenli Li et al.** [19] proposed stochastic dynamic level scheduling (SDLS) algorithm to minimize the makespan. This algorithm outperforms when tasks arrive randomly.

**Jorge et al.** [20] proposed parallel heterogeneous earliest finish time (P-HEFT) which is an extension to HEFT. P-HEFT supports parallel task DAG which provides optimized makespan that makes it suitable for image processing type of application.

**Pravanjan Choudhury et al.** [21] proposed online scheduling of dynamic task graphs. Algorithm provides dynamic path selection option by scheduling tasks at run time. The proposed algorithm is assumed to be limited to homogeneous systems. But it can be extended further to heterogeneous systems by taking the base of this algorithm.

**Tang, Z. et al.** [22] proposed self-adaptive reduce scheduling (SARS) for Hadoop platform. During MapReduce phase, it reduces the waiting time by selecting an adaptive time to schedule the reduce task. This method reduces the turnaround time.

# 5 Conclusion and Scope for Future Work

Several scheduling algorithms have been developed for scheduling in heterogeneous distributed system. The paper reviews various scheduling algorithms along with its merits and possible future enhancements. It is found that each algorithm can be further improvised for better performance. This study is primarily focused on parameters such as scheduling type, resource utilization, workload balancing, efficiency, speed optimization, and high performance. Some of the algorithms discussed in this paper, emphasizes on static scheduling. These algorithms can be improvised and implemented for dynamic scheduling as well, which can be considered as scope for future research.

# References

1. Topcuoglu, H., Hariri, S., Wu, Min-You: Performance-effective and low-complexity task scheduling for heterogeneous computing. IEEE Trans. Parallel Distrib. Syst. **13**, 260–274 (2002)
2. Padole, M.: Distributed computing for structured storage, retrieval and processing of DNA sequencing data. Int. J. Internet Web Technol. **38**, 1113–1118 (2013)
3. Foster, I., Kesselman, C.: The Grid. Morgan Kaufmann, Amsterdam (2004)

4. Yuxiong, H., Liu, J., Hongyang, S.: Scheduling functionally heterogeneous systems with utilization balancing. IEEE Int. Parallel Distrib. Process. Symp. 1187–1198 (2011)
5. Zhu, Y: A survey on grid scheduling systems, Department of Computer Science, Hong Kong University of science and Technology (2003)
6. EL-Rewini, H., Lewis, T., Ali, H.: Task scheduling in parallel and distributed systems. Prentice Hall, Englewood Cliffs, N.J. (1994)
7. Casavant, T., Kuhl, J.: A taxonomy of scheduling in general-purpose distributed computing systems. IEEE Trans. Softw. Eng. **14**, 141–154 (1988)
8. Zheng, W., Sakellariou, R.: Stochastic DAG scheduling using a Monte Carlo approach. J. Parallel Distrib. Comput. **73**, 1673–1689 (2013)
9. Munir, E., Mohsin, S., Hussain, A., Nisar, M., Ali, S.: SDBATS: a Novel algorithm for Task scheduling in heterogeneous computing systems. Parallel and Distributed Processing Symposium Workshops & PhD Forum (IPDPSW), 2013 IEEE 27th International. 43–53 (2013)
10. Kwok, Y., Ahmad, I.: Static scheduling algorithms for allocating directed task graphs to multiprocessors. ACM Comput. Surv. **31**, 406–471 (1999)
11. Kanemitsu, H., Hanada, M., Nakazato, H.: Clustering-based task scheduling in a large number of heterogeneous processors. IEEE Trans. Parallel Distrib. Syst. **27**, 3144–3157 (2016)
12. Abdelkader, D., Omara, F.: Dynamic task scheduling algorithm with load balancing for heterogeneous computing system. Egypt. Inform. J. **13**, 135–145 (2012)
13. Wang, G., Wang, Y., Liu, H., Guo, H.: HSIP: a novel task scheduling algorithm for heterogeneous computing. Sci. Progr. **2016**, 1–11 (2016)
14. Munir, E., Ahmad, S., Nisar, W.: PEGA: a performance effective genetic algorithm for task scheduling in heterogeneous systems. In: High Performance Computing and Communication & 2012 IEEE 9th International Conference on Embedded Software and Systems (HPCC-ICESS), 2012 IEEE 14th International Conference. 1082–1087 (2013)
15. Ahmad, S., Liew, C., Munir, E., Ang, T., Khan, S.: A hybrid genetic algorithm for optimization of scheduling workflow applications in heterogeneous computing systems. J. Parallel Distrib. Comput. **87**, 80–90 (2016)
16. Cardellini, V., Grassi, V., Presti, F., Nardelli, M.: Distributed QoS-aware scheduling in storm. DEBS'15 Proceedings of the 9th ACM International Conference on Distributed Event-Based Systems. 344–347 (2015)
17. Arabnejad, H., Barbosa, J.: List scheduling algorithm for heterogeneous systems by an optimistic cost table. IEEE Trans. Parallel Distrib. Syst. **25**, 682–694 (2014)
18. Khaldi, D., Jouvelot, P., Ancourt, C.: Parallelizing with BDSC, a resource-constrained scheduling algorithm for shared and distributed memory systems. Parallel Comput. **41**, 66–89 (2015)
19. Li, K., Tang, X., Veeravalli, B., Li, K.: Scheduling precedence constrained stochastic tasks on heterogeneous cluster systems. IEEE Trans. Comput. **64**, 191–204 (2015)
20. Barbosa, J., Moreira, B.: Dynamic scheduling of a batch of parallel task jobs on heterogeneous clusters. Parallel Comput. **37**, 428–438 (2011)
21. Choudhury, P., Chakrabarti, P., Kumar, R.: Online scheduling of dynamic task graphs with communication and contention for multiprocessors. IEEE Trans. Parallel Distrib. Syst. **23**, 126–133 (2012)
22. Tang, Z., Jiang, L., Zhou, J., Li, K., Li, K.: A self-adaptive scheduling algorithm for reduce start time. Future Generation Computer Systems **43–44**, 51–60 (2015)

# Part III
# Advanced Computing: Knowledge Representation

# Knowledge Reasoning for Decision Support (KRDS) in Healthcare Environment

Nawaf Alharbe, Lizong Zhang and Anthony S. Atkins

**Abstract** This research develops a Knowledge Transformation for Knowledge Reasoning for Decision Support (KRDS) that combines information from domain experts and sensor data, to provide integrated decision support with real-time capability. The paper relates the processes of Knowledge Acquisition (KA) and Knowledge Harvesting (KH) within Knowledge Management to the healthcare environment where sensors are used to track and monitor 'objects' (patient, staff and assets) which can be transformed into a Smart Hospital Management Information System (SHMIS). A case study based on Medina Maternity and Children Hospital (MMCH) in Saudi Arabia is used to outline the proposed transformation.

**Keywords** RFID · ZigBee · KMS · KA · KH · KRDS · KM · CoPs and SHMIS

## 1 Knowledge Acquisition

Knowledge Acquisition (KA) is the process of acquiring knowledge from a human expert, or a group of experts, known as Community of Practice (CoPs), and using this knowledge to build knowledge-based systems [1]. Knowledge acquisition has been defined as the knowledge that is acquired by both informal and formal

N. Alharbe (✉)
College of Community, Taibah University, Badr, Kingdom of Saudi Arabia
e-mail: nrharbe@taibahu.edu.sa

L. Zhang
School of Computer Science & Engineering, University of Electronic Science
and Technology of China, No. 4, Section 2, North Jianshe Road,
Chengdu 610054, Sichuan, China
e-mail: l.zhang@uestc.edu.cn

A.S. Atkins
Faculty of Computing, Engineering and Sciences, Staffordshire University,
Beaconside, Stafford ST18 0AD, UK
e-mail: a.s.atkins@staffs.ac.uk

© Springer Nature Singapore Pte Ltd. 2018
R.K. Choudhary et al. (eds.), *Advanced Computing
and Communication Technologies*, Advances in Intelligent Systems
and Computing 562, https://doi.org/10.1007/978-981-10-4603-2_13

processes, and serves as the basis for decision-making [2]. KA is an experience-built process to develop methods and tools that make the task of capturing and validating an expert's knowledge as efficient and effective as possible [3]. KA has also been defined as accepting knowledge from the external environment and transforming this knowledge so that it can be used by an organisation [4, 5]. KA components at organisational level contain the activities of extracting, interpreting and transferring knowledge, so as to improve existing knowledge [4].

The acquisition of knowledge may be rewarding work, allowing knowledge exploitation for competitive advantage by means of new system development stages, technological distinctiveness and sales rate efficiency [6]. However, KA may also be challenging, because it is difficult to get expert staff in the domain, technology vendors and knowledge engineers to speak the same language and understand concepts in the same way [7]. In certain circumstances, they may not agree, which could result in waste of effort, increased cost and time delays in the acquisition of knowledge that reflects the spectrum of the service provided [6].

## 1.1 Knowledge Acquisition in Health Care

KA is required to create effective Smart healthcare systems [8]. Typically, KA could be accomplished with experts such as doctors, nurses and other stakeholders such as senior administrative and managerial staff. In a hospital environment, the applications of KA and KM should include more efficient patient systems and increased security. In the context of the present study of healthcare systems in Saudi Arabia, KA is fundamental in supporting the implementation of a knowledge reasoning layer to support the development of a Smart Hospital Management Information System (SHMIS), which in turn, should lead to better management of resources and improved services for patients. Hanson and Kararach noted that a healthcare organisation needs to transfer the right knowledge to the right people at the right time [9].

However, working with KA in a healthcare environment presents a number of challenges. A great amount of knowledge is tacit and is not directly accessible. It needs to be converted to make it explicit [10]. Some experts find it difficult to communicate the knowledge as required by the stakeholder [2]. It is also possible that a healthcare organisation would place restrictions on experts in order to reduce the transmission of knowledge to the competitors.

## 1.2 The Stages of KA

Holsapple and Joshi, define KA as knowledge that is first identified, then captured, and organised before being transferred [11]. These four stages are outlined in the following:

- *Identifying*: Appropriate knowledge within the organisation's existing resources. This includes locating, accessing, valuing, and/or filtering knowledge.
- *Capturing*: Identified knowledge from outside. This involves extracting. Collecting, and/or gathering knowledge, which is considered valid and useful.
- *Organising*: Captured knowledge. This involves distilling, refining, orienting, interpreting, packaging, assembling. And/or transforming captured knowledge into representations that can be understood and processed by another knowledge manipulation activity.
- *Transferring*: Organising knowledge. This involves communication channel identification and selection, scheduling, and sending. This transfer can be an activity that immediately uses the knowledge or internalises it for subsequent use.

## 1.3  Relationship Between Knowledge Acquisition and Knowledge Harvesting

In this study, Knowledge Acquisition and Knowledge Harvesting (KH) are distinguished and concepts from both KA and KH are utilised. The first three stages of KA, as identified in 1.2, are regarded as the KA stage of the KRDS. The final stage of KA [11] is the transfer of knowledge to an activity, or as in healthcare, to an actor, who can make use of this knowledge. In this paper, the knowledge transfer stage, is discussed in more detail in Sect. 2, and is covered by the stages of the Knowledge Harvesting process. KA is understood in this paper, to cover the acquisition of both tacit and explicit knowledge and provides the input for decision-making. Advances in digital technology, and particularly smart devices such as 'object tags', mean that we also need to consider the process whereby, for example, sensor data collected in real time may become part of knowledge management strategies.

As discussed in the following sections, Knowledge Harvesting is understood here to be a wider and more dynamic concept than knowledge acquisition. The definition of KH, discussed in Sect. 2, like the definition of KA, emphasises the importance of knowledge capture and the transformation of implicit to explicit knowledge, but also emphasises the organisational transformations that should be made and the iterative nature of knowledge harvesting. Thus, KA is seen as one of the stages of KH, and as an enriched traditional KA with automated data collection of, for example, sensor data. This research aims to link both KA and KH to the SECI model, to show how knowledge is to be created and shared between the users in the healthcare environment.

Although, KA is an older and structured approach, which gives guidance on how information is extracted and combined, it is limited, in that it does not place sufficient emphasis on the transformation stage. KH, on the other hand, is a new and much more structured approach, and places greater emphasis on the active process of transformation. Furthermore, KH focuses on 'packaging', finding the right format, and applying and continually evaluating the information. This research uses a combination of both KA and KH, which extends and evaluates the information from both the CoPs and sensor devices in order to transform this information into a usable format.

## 2 Knowledge Harvesting

Knowledge Harvesting (KH) is a structured, results-driven process for capturing vital knowledge, including deep insight and complex cognitive processes [12]. Knowledge production and idea generation are important elements in creating a long-term competitive advantage. Knowledge Harvesting is regarded as a fundamental part of Knowledge Management [9]. KH can be used to help hospitals uncover and capture vital knowledge that enables the hospital to make more efficient use of resource, to achieve maximum benefit for patients, and to provide more control provisions in the hospital, so as to raise the level of performance [13]. Through the harvesting of knowledge and the transformation of Knowledge for Decision Support, KH could provide healthcare intelligence, allowing 'what-if' scenarios and real-time decision-making support.

## 2.1 The Stages of Knowledge Harvesting

These eight stages of Knowledge Harvesting are outlined in the following [12, 14]:

- **Focus**: to help the users to identify the knowledge in their organisation that is most urgent and important, and to select and prioritise the processes or strategies for harvesting.
- **Find**: to provide guidance to the users on locating experts and existing support information.
- **Elicit**: to show the users how to conduct effective harvesting sessions.
- **Organise**: to instruct users on how to make sense of the information gathered via interviews and documents. In addition, to instruct users on how to identify patterns and organise the knowledge into logical support information and guidance.
- **Package**: to instruct users on how to determine the best vehicle for packaging the knowledge, and to select the most appropriate medium for packaging and sharing the knowledge assets and transfer them to other users.
- **Apply**: to show users how best to apply know-how, by accessing the knowledge assets at the point of need, and using the know-how in their regular work.
- **Evaluate**: to provide a feedback mechanism for evaluation as a continuous process. Also, to provide tools and guidance for measuring the effectiveness of

the knowledge asses. Healthcare organisations should also consider the value of know-how over time. Knowledge that is codified in static documents can date quickly but should be kept up-to-date, relevant and as small as possible [15].

- **Adapt**: As a result of the ongoing appraisal and feedback, to show users how to make the necessary alterations.

A key feature of knowledge harvesting is a collection of methods for eliciting information about four types of knowledge: declarative knowledge, procedural knowledge, contextual knowledge and social knowledge.

## 2.2 Knowledge Harvesting with Sensor Data in Healthcare

There is no set method for knowledge harvesting in the healthcare environment, although rules are available that facilitate knowledge acquisition, such as one-to-one, face-to-face interviews with hospital experts, and extracting knowledge from the organisation's database, as the automatic system records specific situations in real time. Furthermore, Smart systems could easily apply specific know-how by utilising the harvested knowledge for the purpose of developing, improving and training hospital staff [13]. KH could be used in any field of human activity, and particularly in management [9]. There are numerous benefits of sharing knowledge stocks between experts [9]. This may apply in healthcare environments, as the harvested knowledge could be made available with the requirement to memorise it. KH is one way of capturing the knowledge available within different organisations [15]. KH could identify and extract critical knowledge, and then capture, organise and disseminate it to the healthcare staff [16]. This suggests that using KH in a healthcare environment could reduce the scope for human error.

## 3 The Proposed Knowledge Reasoning for Decision Support (KRDS) Model

The proposed KRDS model builds upon the concepts of KA and KH and is a novel approach to knowledge harvesting, acquiring and transforming expert knowledge. Within a hospital environment, this is to be achieved by incorporating sensor data into the Smart Hospital Management Information System (SHMIS) to improve future operational and staff performance. The focus is on using knowledge for system transformation, in order to develop a smart hospital system and on evaluating and refining the transformation process.

## 3.1    The Stages of KRDS

It is important to transfer the right knowledge to the right people at the right time
and to help them apply it, if we are to improve the quality of care delivery [15]. The
five stages of this process are outlined in Fig. 1 and are further discussed as follows:

**Stage 1: Knowledge Acquisition**

The first stage of KRDS builds upon traditional knowledge acquisition approaches.
In this stage, knowledge is acquired through discussions and meetings with professional staff in the healthcare organisation, together with CoPs, in order to share
and improve existing knowledge. The initial step involves determining the top
performing people and their critical activities. The knowledge acquisition stage
includes eliciting tacit and explicit knowledge from human experts and documen-
tary sources. After identifying the issues and the right experts and activities, a
comprehensive understanding of these activities is elicited from the experts.

**Stage 2: Identification**

Users determine how best to use the knowledge that has been gained and identify
application areas. This stage considers what data could be obtained from sensors
and how this data could be used to supplement or replace non-sensor data. Tools
and measures are identified together with targets and possible transformations.

**Stage 3: Contextualisation**

In this stage, the acquired knowledge is analysed and interpreted over several
meetings of the extended CoPs (experts, supervisors, vendors and clinical staff)

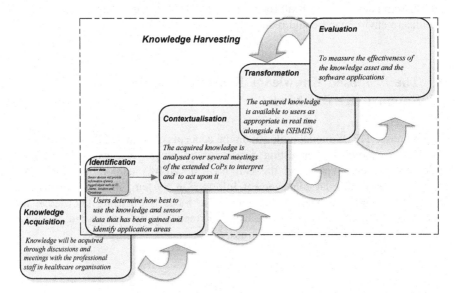

**Fig. 1** A conceptual view of the proposed KRDS

who then act upon it. This stage of the KRDS has a similar role to the Organise and Elicit stage in the Knowledge Harvesting model.

**Stage 4: Transformation**

The transformation stages of the KRDS covers the Package and Apply stages of the KH model. Data is extracted from a real-time object movement as recorded by the Smart Hospital Information Management System (SHIMS) database, which is exchanged between the stakeholders in different locations for different purposes. In addition, this will enable the automatic construction of large knowledge hubs. To illustrate the KRDS, Fig. 1 gives an outline of the process as applied to transform paper-based hospital admission processes into a Smart admission processes and demonstrates how the different stages of the KRDS map to a real-world application.

**Stage 5: Evaluation**

The purpose of this stage is to measure and evaluate the relevance and effectiveness of the knowledge gathered from experts and from software applications (sensor devices). In addition, this process should be performed consistently so that the database is kept up to date, relevant and as small as possible. The accuracy of the harvested knowledge is tested in order to ensure that it is fit for purpose before it is transferred to the targeted end-user for use in the Application stage. In the event of errors, the process would restart, in order to ensure the quality of the information. Figure 1 shows a feedback loop to the contextualisation stage, so that the model provides continuous feedback and interaction to ensure relevance and consistency and to support transformation.

# 4  Case Study—Monitoring of Medical Equipment at Medina Maternity and Children's Hospital (MMCH)

The following case study focus on using a KRDS process combined with sensor data to improve location of patients and staff and maternity security. Based on the results from these studies, the information received from the RFID tags will be used to propose a new Smart System for Monitoring Mobile Medical Equipment. A passive RFID tag and/or ZigBee tag will be attached to all medical equipment. RFID readers and one ZigBee coordinator are fixed into the wards' entrance 'gate', and a ZigBee reader will be deployed in the ceiling in the covered area The KRDS model, as shown in Fig. 1, shows how the expert knowledge within a hospital environment by incorporating sensor technology can be used to transform and improve future operational and staff performance.

## 4.1 Case Study—Overview

In 2009 a study by GS1 UK and Nursing Times [17] indicated that nurses waste time in unproductive activities, showing that more than 33% of nurses spent at least an hour trying to locate medical equipment, such as pumps and drip stands, during an average hospital shift. This is equivalent to 40 h a month and costs the NHS nearly £900 million per year. Medina Maternity and Children's Hospital (MMCH) in Saudi Arabia is facing a similar challenge as to how to effectively manage and monitor the location of essential mobile medical equipment, such as beds, wheelchairs, defibrillators, pumps, incubators, syringes and feeding devices [18]. These constitute important hospital assets and require appropriate management in order to help ensure constant supply and to provide quality patient care. Using nurse time to locate equipment is seen as an unproductive use of nursing resources. Active RFID technology has already been used for asset tracking in Addenbrooke hospital and Derby Hospitals NHS Trust in the UK, [19]. In the current research, passive RFID tags were used, partly on grounds of cost but also because this system uses a combination of passive RFID and ZigBee tags for objects identified as having high vulnerability. This study investigated whether as part of the proposed transformation to a Smart Hospital System, mobile medical equipment could be tagged and tracked.

## 4.2 Knowledge Acquisition

In the study of the proposed Smart monitoring of mobile medical equipment, the first stage was knowledge acquisition from the expert professionals (doctors, nurses, senior administrators, medical engineers and IT staff) who are involved in the monitoring of mobile medical equipment. Several meetings were held at the Medina hospital with the expert professionals, to describe and understand the conventional process and its possible transformation. Discussions were conducted at MMCH concerning the type of equipment in use on the wards, its purpose, which type was in most demand, and how the supply of equipment was managed to meet specific patients' needs, etc. This also included a discussion on the cost of the service for mobile medical equipment and of maintenance schedules. During the meeting, the medical staff identified a need to improve security for mobile medical equipment as some of it is very expensive.

## 4.3 Identification

The second stage involved meetings with doctors, nurses, senior administrators, medical engineers, IT staff and the Smart system developer to discuss needs and requirements. The meetings developed the specification for the technical equipment, and the location and devices required. Passive RFID combined with ZigBee technology were selected in order to provide information on the location of all tagged mobile medical equipment. This information will be analysed, to enable the hospital staff to make competent decisions regarding better use of medical equipment, and meet the healthcare standards.

## 4.4 Contextualisation

In this stage, the acquired knowledge concerning medical equipment was analysed and interpreted over several meetings of the extended CoPs (doctors, nurses, senior administrators, medical engineers, IT staff and the SHMIS developer). Using both RFID and/or ZigBee technologies will allow accurate location of the equipment; identifying who was using it last time, and producing tracking reports and information about any forthcoming maintenance schedules. The overall cost of a ZigBee (£100) and RFID passive tag (<5p) is relatively low in relation to the cost of the medical equipment, for example Ultra sound equipment cost approximately £60,000. Missing this equipment could result in equipment becoming unusable, which could then be damaged and needed to be replaced.

## 4.5 Transformation to Smart System of Monitoring Mobile Medical Equipment

In Medina Maternity and Children's Hospital (MMCH), as in the rest of the Saudi hospital system, there is no system for monitoring and tracking mobile medical equipment. This stage of the proposed Smart monitoring of mobile medical equipment will require all the equipment to have identification bands, including RFID tag and/or ZigBee tag. The proposed SMHIS would be able to record in real time the time and location at which equipment arrives and leaves a ward. A passive RFID tag and/or ZigBee tag will be attached to all medical equipment. RFID readers and one ZigBee coordinator are fixed into the wards' entrance 'gate', and a ZigBee reader will be deployed in the ceiling in the covered area. As the equipment moves between hospital wards, it is automatically detected by the readers. This data will be sent to the database, using the hospital network. The database is updated each time this equipment moves, or leaves and enters a ward or department, and the system user will be able to visualise the information, including location and movement history as well as maintenance schedules and the purpose of using this equipment in relation to the patient. The transformation to a Smart system for equipment tracking and location will help the hospital to make decisions regarding planned services, and to assess movement planning in order to optimise utilisation.

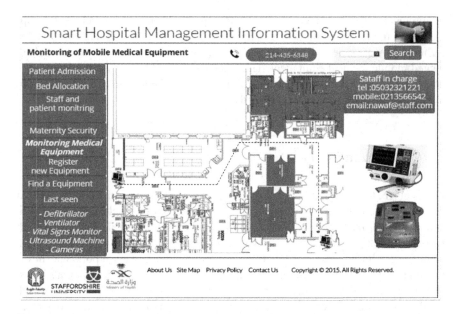

**Fig. 2** Proposed smart monitoring for mobile medical equipment at MMCH

In the event of unexpected or unexplained movement of mobile medical equipment, the proposed system would send an alert to the nearest available member of staff. Figure 2 shows the visualisation of data that could be used by hospital staff.

In order to register new equipment, the new system will require functionality to create a record for the equipment purchased. This record includes all the relevant information, such as production year, maintenance schedule or battery level, as well as the unique RFID and/or ZigBee identifier. The system will also allow users to track and locate registered equipment. This visualisation data will be clearly presented on the hospital floor layout, to allow quick identification and location in order to make Smart decisions. Once the equipment is taken away from the hospital, the system will keep the records for future use. If ZigBee tags are used, they will be removed from the equipment before disposal. The tags in case of ZigBee sensor could be reused when registering new equipment [20].

## 4.6 Discussion

The proposed smart system for monitoring mobile medical equipment, as part of the KRDS process, was developed, in the laboratory environment and tested in relation to the functionality of the system by tagging several trolleys. The identification bands containing passive RFID tags and a separate ZigBee band were used for tagging the equipment. The RFID and ZigBee test equipment was installed in the research centre area, in order to test the efficiency of the proposed new system and to identify and prevent anomalies. The security application in the test laboratory operated successfully and triggered an alarm to notify the system developer when the objects left or re-entered the laboratory.

# 5 Conclusion

In a healthcare environment, KRDS supports the transfer of knowledge from domain experts to support decision-making, and the transformation to a Smart Hospital Management Information System (SHMIS). Competitive health organisations should be increasingly flexible and adaptable to new technology and techniques. This would require the development of the current strategies, which must interact and be commensurate with the changes that would be incurred. To address these needs, the tacit knowledge and know-how of experts and top performers in any organisation need to be captured and documented, to make the arduous task of capturing and validating it efficient and effective. Knowledge acquisition and harvesting are new techniques particularly in the healthcare sector, which has yet to find valid and reliable tools to mine the significant knowledge. The proposed process is important in getting the right knowledge to the right people at the right time, and helps them apply it, so that the quality of care delivery is improved. The aim is to create, support, capture, store and share knowledge in the most efficient and timely manner. Much of the functionality in a hospital environment needs to be linked to the system so as to enhance patient care.

# References

1. Kendal, S., Creen, M.: Knowledge acquisition. In: An Introduction To Knowledge Engineering, pp. 89–107. Springer, London (2007). Available from: http://www.springerlink.com/index/10.1007/978-1-84628-667-4. Accessed 25 May 2016
2. Warren, L.E., Mendlinger, S.E., Corso, K.A., Greenberg, C.C.: A model of knowledge acquisition in early stage breast cancer patients. Breast J. 18(1), 69–72 (2013)
3. Iria, J.: A core ontology of knowledge acquisition. Semant. Web: Res. Appl. 233–247 (2009)
4. Liao, S., Wu, C., Hu, D., Tsuei, G.: Knowledge acquisition, absorptive capacity, and innovation capability: an empirical study of Taiwan's knowledge-intensive industries. Technology, pp 160–167 (2009)
5. Pacharapha, T., Vathanophas Ractham, V.: Knowledge acquisition: the roles of perceived value of knowledge content and source. J. Knowl. Manage. 16(5), 724–739 (2012)
6. Yli-Renko, H., Autio, E., Sapienza, H.J.: Social capital, knowledge acquisition, and knowledge exploitation in young technology-based firms. Strategic Manage. J. 22(6–7), 587–613 (2001)
7. Musen, M.A.: The knowledge acquisition workshops: a remarkable convergence of ideas. Int. J. Human-Comput. Stud. 71(2), 195–199 (2013)

8. Moncur, W., Mahamood, S., Reiter, E., Freer, Y.: Involving healthcare consumers in knowledge acquisition for virtual healthcare. Paper presented at virtual healthcare interaction 2009 (AAAI Fall Symposium Series), Washington D.C., United States (2009)

9. Hanson, K., Kararach, G.: The challenges of knowledge harvesting and the promotion of sustainable development for the achievement of the MDGs in Africa. In: Harare: African Capacity Building Foundation (2011). Available from: http://elibrary.acbfpact.org/collect/acbf/index/assoc/HASH016a.dir/doc.pdf. Accessed 12 May 2016

10. Birk, A., Surmann, D., Althoff, K. (1999). Applications of knowledge acquisition in experimental software engineering. In: Fensel D., Studer R. (eds.) Knowledge Acquisition, Modeling and Management. Lecture Notes in Computer Science, pp. 67–84. Springer, Berlin

11. Holsapple, C., Joshi, K.: A knowledge management technology. In: Holsapple, C.W. (ed.) Handbook on Knowledge Management 1. Springer, Berlin (2004)

12. Wilson, L.T., Wilson, H.L.: Local knowledge harvesting: a method for leveraging knowledge to grow economies in under-resourced areas. In: 2011 International Conference on Knowledge Economy (ICKE), 2nd International Conference on Knowledge Economy East London, Eastern Cape Province, Republic of South Africa. 2011, pp. 1–35 (2011)

13. Servin, G., Brun, C. De.: ABC of knowledge management. NHS National Library for Health: Specialist Library (2005). Available from: http://www.fao.org/fileadmin/user_upload/knowledge/docs/ABC_of_KM.pdf. Accessed 20 Nov 2015

14. Frappaolo, C., Wilson, L.T.: Implicit knowledge management: the new frontier of corporate capability [Electronic version], pp. 1–7. Knowledge Harvesting, Inc. (1999). Available from: http://www.knowledgeharvesting.com/documents/ ImplicitKM.pdf. Accessed 25 Nov 2015

15. Serrat, O.: Harvesting Knowledge. Asian Development Bank, Washington, DC (2010)

16. Khursani, S., Buzuhair, O., Khan, M.: Strategy for rapid transformation of Saudi Arabia by leveraging intellectual capital and knowledge management. Saudi Aramco J. Technol. (2011). Available from: http://rushkhan.org/uploads/3/2/4/5/3245782/sa_intellectual_capital.pdf. Accessed 2 Feb 2016

17. GS1, UK.: Nurses lose up to one quarter of their working day hunting for medical items (2009). Available from: http://www.gs1uk.org/news/Pages/PressReleaseDetails.aspx?PRID=5. Accessed 29 Jan 2016

18. Alharbe & Atkins.: Private communication—Madina Maternity and Children Hospital (2015)

19. Swedberg, B.C.: RFID boosts medical equipment usage at UK Hospital, pp. 3–5 (2013)

20. Atkins, A., Zhang, L., Yu, H.: Application of RFID Technology in e-Health Management and Outsourcing in Bhutan. In: 5th IEEE International Conference Advance Computer Communication Technology (ICACCT-2011), pp. 779–784. APIIT, Haryana, India 2011

# Object Detection from Video Sequences Using Deep Learning: An Overview

Dweepna Garg and Ketan Kotecha

**Abstract** One of the challenging topics in the field of computer vision is the detection of the stationary/non-stationary objects from a video sequence. The outcome of detection, tracking, and learning must be free from ambiguity. For effectively detecting the moving object, first the background information from the video should be subtracted. However, in the high-definition video, modeling techniques suffer from high computation and memory cost which may lead to a decrease in performance measure such as accuracy and efficiency in identifying the object accurately. It is important to identify the definite structure from a large amount of unstructured data which is a prerequisite problem to be solved. The task of finding the structure from a large amount of data is achieved using Deep Learning 'which is about learning multiple levels of representation and abstraction that help to make sense of data such as images, sound, and text'. The purpose of the paper is to survey the method with which the objects can be efficiently detected from any given video sequence along with the preferable use of the deep learning library.

**Keywords** Deep learning · Object detection · Video sequence · Graphics processing unit · Tensor flow

## 1 Introduction

'Learning'—an eight letter word means that one understands something that 'we' have understood all our life, but in a unique manner. 'Machine' when combined with 'Learning', refers to mainly three terms: task, performance, and experience. It can be framed as the performance that can be measured for a task and this performance

D. Garg (✉) · K. Kotecha
Parul University, Limda, Vadodara, India
e-mail: dweeps1989@gmail.com

K. Kotecha
e-mail: provost@paruluniversity.ac.in

© Springer Nature Singapore Pte Ltd. 2018
R.K. Choudhary et al. (eds.), *Advanced Computing
and Communication Technologies*, Advances in Intelligent Systems
and Computing 562, https://doi.org/10.1007/978-981-10-4603-2_14

measure can be improved with some experience. This concept is referred to as Machine learning [1], a term coined by Arthur Samuel which can be stated as a way to make the computer intelligent. He stated in way back 1959 that machine learning is the field of study that gives the computers the ability to learn without being explicitly programmed. In 1997, Mitchell [2] defined machine learning as 'a computer program that could learn from experience E with respect to some task T and some performance measure P, if its performance on T, as measured by P, improves with experience E.' Machine learning aims to bring a degree of order to the zoo of machine learning problems which is spread to a vast range of applications. The emerging area of Machine Learning research is Deep Learning which is introduced with the motive of moving the machine learning a step closer to artificial intelligence. It is becoming popular mainly because of the following three reasons: First 'drastically increased chip processing abilities', e.g., General Purpose Graphics Processing Units (GPGPUs), second due to its usage of an increased size of data used for training, and third because of the recent advances in machine learning and processing information. 'The above reasons have made the deep learning methods to compute various complex problems and to effectively make use of both unlabelled and labelled data' [3].

The paper attempts to provide an overview of detecting either non stationary or stationary objects accurately from the given video sequence using this emerging machine learning methodology. Furthermore, the purpose of the paper is limited in addressing to the research issues which can be targeted to do the above mentioned methodology. The rest of the paper is organized as follows—Sect. 2 deals with the basic structure. Section 3 targets the research issues mainly focusing on detecting the objects. A brief review of deep learning is presented in Sect. 4. Section 5 highlights the main topic of this paper giving an insight of how the work can be carried out. Section 6 describes the deep learning frameworks useful from programming point of view, Sect. 7 of the paper focuses on the applications of deep learning. Finally conclusions are drawn in Sect. 8.

## 2 The Basic Structure

In Fig. 1, the set of training examples are fed as an input to the learning algorithm. The choice of the learning example depends solely on the user. Hypothesis ($h$) is calculated taking into consideration the size of the house and the learning algorithm. Size of the house is one of the input features considered as an example. The formula for calculating the hypothesis is

$$h(X) = \sum_{i=1}^{n} \theta i X i \tag{1}$$

**Fig. 1** Basic structure of
learning algorithm

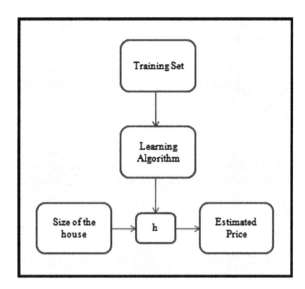

Here $n$ denotes the number of training examples, $\theta$ is the parameter, and $X$ is the input feature. In the above example the size of the house is one of the input features. Then the value of estimated price of the house is calculated. A model is prepared through a training process where it is required to make the predictions and is corrected when those predictions are wrong. The training process continues until the model achieves a desired level of accuracy. It is calculated using $h(X) - y$. Here $h(X)$ indicates the calculated hypothesis value and y indicates the corrected output value.

## 3 Research Issues

Most of the machine learning problems is arising during its applications to real-world problems. Poor performance was observed not because of the choice of learning algorithm but due to the selected training data. The problems faced by training data was—the training data selected was insufficient, training set of data was too small to learn a generalize model or the data contained noise. Due to an increased amount of data, the performance measure (for example, accuracy) was one of the main issues in machine learning. The time taken to process large amount of data was significantly high. To mimic the brain for representing information with great efficiency and robustness is a core challenge in the research of artificial intelligence. Around in 2010, the real impact of machine learning in industry began in large-scale speech recognition [3]. Artificial neural networks were involved in major of successful deep learning methods. In the context of computer vision, object detection, identification of the image of the object in any image or in a video sequence is one of the most important fields which. It is discussed in detail later in the paper.

# 4 Deep Learning

It is also known as deep structured learning or hierarchical learning or deep machine learning. Geoffrey Hinton coined the term 'deep learning'. It is a large family consisting of both supervised and unsupervised learning methods. The computational models are made up of multiple processing layers and each layer learns the representation of data at different abstraction level. Artificial neural network is largely inspired by the way the human brain works. And today a new field has emerged with respect to the research of neural network and it is Deep Learning. Deep learning in collaboration with other algorithms helps in classification [4], clustering, and prediction [5]. It first reads the structure automatically and then when the deep learning algorithm trains it is able to make the guess against the training set and try to bring the guess close to the accurate answer. This is referred to as optimization [6]. It is able to learn the complicated patterns from a large amount of data by combining the computational power and special types of neural network. The word 'deep' can be used to call the network having multiple hidden layers. Adding more layers add to the difficulty in updating weights as the signals [3] propagating becomes weak. Therefore, the weight of the network becomes off the track and it becomes impossible to parameterize a 'deep' neural network with backpropagation. Here 'deep learning' comes into play which helps to train the 'deep' structures of neural network.

With deep learning one can classify, cluster, and predict about the data consisting of video, images, text, time series, and many more. It is mainly used in object detection (face detection, pedestrian detection), visual recognition of objects (ImageNet), speech recognition (conversion from speech to words), and predicting the drug activity. Deep learning [7] has already made a successful advancement in voice search on smart phones and text-to-speech conversion and speech-to-speech conversion. GPU's (graphics processing units) have greatly contributed in image segmentation, object detection, pattern recognition, and natural language processing. GPU's train the deep neural network using large training sets but in a less time. Its computing provides the efficient and parallel computation of large amount of training data. As the GPU takes less time to compute the large amount of data and works effectively well for machine learning, it is believed that the graphics processing unit can accelerate the machine learning algorithms very well. The reason for widely using GPU is its computational power and its capabilities which are growing at a faster rate as compared to CPU. The deadly combination of CPU and GPU can produce the best value in terms of computation power, performance, and price.

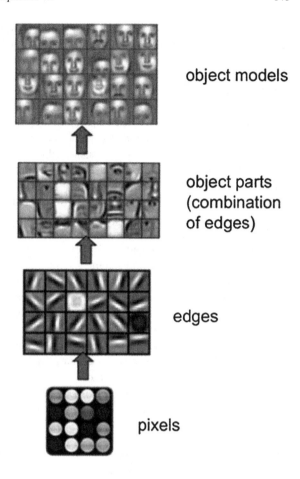

object models

object parts
(combination
of edges)

edges

pixels

Deep Learning uses backpropagation algorithm in order to depict how the internal parameters are changed by the machine in order to compute the representation in each layer with respect to its previous layer. Initially the pixels are converted to the edges, combination of edges makes the object parts and finally the object model is prepared as depicted in Fig. 2.

Convolutional neural networks have found to succeed in a variety of tasks of computer vision especially focusing on the recognition. The concept of optical flow [8] was introduced keeping in mind the relative motion of both the viewer and the object. It highlights the important information about the spatial arrangement of the viewed object along with its displacement. It is regarded as one of the vital ingredient in solving many complex computer vision tasks. The main focus of estimating optical flow is to find per-pixel localization and then mapping the difference between the two images which are fed as inputs to the system. For this, the feature representations are to be learnt. By using various optical flow algorithms, the difference between the frames can be calculated accurately. Evaluating these algorithms does consume time but it has shown great improvement in training the deep convolutional neural network. Computation of optical flow cannot be done for a single point in the image. Rather, the neighboring point should also be considered as each image point has two velocity fields which may change due to the change in any one of the field.

Optical flow [9] estimation has been one where CNN have found to succeed. The motion of any object is perceived when there is a change in the picture. A moving object may produce a constant pattern in brightness. In order to avoid the variations in brightness, the shading surface is considered to be flat and the illumination where the light falls is assumed to be uniform. These conditions help in determining the brightness of the image to be differentiated. Hence the motion of such patterns is determined by viewing the motion of the corresponding points [8]. Calculating the velocities of such points on the object is what known as the optical flow.

Figure 3 illustrates that the information first gets spatially compressed in a con-tractive part of the network and then it gets refined in the expanding part. Hence, it can be said that the 'architecture is trained from end-to-end'. The main idea behind is to exploit the capability of CNN of learning the powerful features at multiple levels of abstraction. It helps in determining the actual difference in the input images based on the features. A huge amount of training dataset is needed to predict the optical flow in training a network. Generation of optical flow from a video sequence is a difficult job and hence a popular synthetic dataset named Flying Chairs dataset [10] has been used. This dataset consists of random background images from Flicker which consists of the image segment of chairs. It is used in training the CNN. For faster processing, implementation on GPU is preferred over CPU.

**Fig. 3** End-to-end supervised learning of optical flow [24]

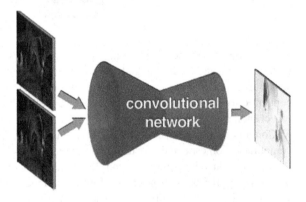

# 5  Object Detection with Deep Learning

It is easier for the humans to detect and recognize the object in the image irrespective of its different viewpoint. Vision begins with the eyes and truly takes place in the brain. So, such cameras are required which can see and understand what humans can see and understand. Cameras take pictures by converting light in 2D pictures. But these are lifeless numbers. They do not carry meaning in themselves. To take pictures does not mean to see. And to see means to understand. For this, the first thing is make the computer understand the object. If the machine is able to detect the objects same as the humans do with great accuracy, then it can lead to wonders in the real world. Another important issue related to object detection is to identify the moving object. Moving object is mainly composed of feature representation [11] and statistical modeling of the scene around. The feature representation takes into consideration the predefined features such as the color code, the texture features, the shape of the feature, and the background against which the object is positioned. The represented features are first learnt using the technique of feature learning where a transformation is carried out from such raw information to a successful representation of an object. It is necessary to discover the required features from the raw input data. Feature learning is also known as the representation learning and is mainly divided in supervised and unsupervised learning [12]. Supervised learning helps in learning the features with the labeled input data. Examples: multi-layer perceptron, neural network. Whereas unsupervised learning the features are learned with the unlabeled input data. Examples include clustering, autoencoders. A broader family of machine learning method named as deep learning is based on learning the representations of the data. *Research can be carried out to effectively represent the feature from a large amount of unstructured data.*

Statistical modeling technique mainly focuses on detecting and tracking [13] the foreground objects (such as car, humans, etc.) from a video sequence by subtracting the background image. *In order to model the foreground image—accuracy in detecting the object, noise and automatic choice of the relevant parts of the object are the key areas of research.* Background subtraction [14] is also known as foreground detection. It is widely used technique for detecting the moving objects [15] from the static cameras. Feature extraction [11], handling deformation and occlusion handling, and classification are four major components in object detection.

After preprocessing the image, the localization of object is required which makes use of this technique. It computes the difference between the current frame and the reference frame, referred to as the background image. It is carried out if the required image is in the video stream. Background subtraction is used in the areas of surveillance and learning of animals, tracking and learning of a specific car on road, human pose estimation, etc. The flow of background subtraction in Fig. 4.

**Fig. 4** Background
subtraction

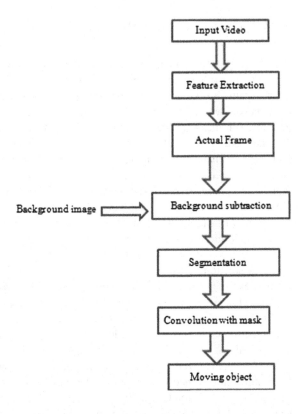

The parameters of study for detecting the moving object from a video sequence can be the accuracy (how accurately the object is identified), efficiency in terms of performance measure and the computational complexity of finding out the desired object.

Figure 5 illustrates the widely used example of object detection from a video sequence where in the vehicles can be monitored, detected and tracked.

The challenges in background subtraction arise with the outdoor scenes, i.e., with respect to rain, wind, different viewpoint of the object or illumination change [16]. In order to detect the object in a better way, it is mandatory to first understand the image. Understanding [17] the image is done using Deep Convolution Network. In this, the inputs are small portions of the image which is fed to the lowest layer in the hierarchical structure. In each layer where information propagates, digital filtering is carried out in order to get the most noticeable features of the observed data. The neurons, also known as the processing units process the features providing a level of invariance to shift, scale, and rotation. Nice megapixels cameras have been developed but yet there is not given vision to the blind.

**Fig. 5** Example depicting
vehicle detection [25]

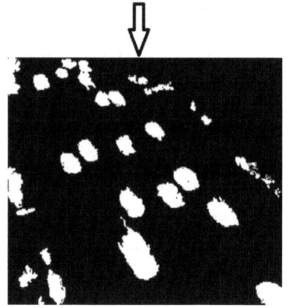

# 6  Deep Learning Frameworks

Some popular frameworks of deep learning are as follows.

**Caffe**: The earliest known mainstream deep learning toolkit was developed by
Berkeley Vision and Learning Center (BVLC) and by community contributors [5].
Caffe has worked well with innovation and application. Caffe can process over 60M
images per day with a single NVIDIA K40 GPU [18]. It is one of the best choices
with respect to deployment because of its cross-platform nature. It is C++ based and

has the ability to be complied on variety of devices. Caffe works well with Convolution Neural Network principles and parts. Mainly the steps to train a CNN with Caffe are: (i) The first step involves data preparation wherein the images are preprocessed and are stored in the format which can be used by Caffe, (ii) The second step moves with model definition where CNN architecture is chosen and the parameters are defined in a configuration file, (iii) Then comes the solver definition where solver holds the responsibility of optimizing the model. The parameters of the same are stored in the configuration file, (iv) The final step is model training wherein the model is trained by executing a Caffe command from the terminal. The trained model is then stored in the file.

**Theano**: Developed by LISA group (now MILA), run by Yoshua Bengio at the University of Montreal [19] is a framework developed by keeping in mind the deep learning algorithms. Theano compiles the program written in Python to efficiently run on GPUs or CPUs. More than a deep learning library, it is a research platform. Theano works well with deep convolutional network, deep belief network and stacked denoising autoencoders.

**Torch**: It was developed by Ronan Collobert, Clement Farabet, and Koray Kavu Cuoglu for research and development in deep learning algorithms. It was promoted by the CILVR lab at New York. Facebook AI lab, Twitter and Google Deepmind later used and further developed this deep learning library. Torch makes use of C/C++ libraries as well as CUDA for GPUs [20].

**TensorFLow**: An open source software library used for machine intelligence is **TensorFlow** [2]. The data flow graphs are used for numerical computation. The mathematical operations are represented by the nodes of the graph and the edges represent the tensors (the multi-dimensional data arrays) to communicate between the nodes. The flexible architecture of TensorFlow allows the users to deploy the computation to one or more central processing unit or the graphics processing unit in a server or desktop or any cellular device with a single API. TensorFlow was originally developed by the engineers working in Google Machine Intelligence Research organization for conducting the research on deep neural networks and machine learning. The system claims the developers is 'general enough to be applicable in a wide variety of other domains as well'.

# 7 Applications

The machine learning technologies are being widely used, e.g., in marketing, finance, telecommunications, and network analysis. One of the well-known examples of machine learning is the concept of web page ranking [21], where the search engine finds the web pages relevant to the query given by the user according to the order of relevance. Another such application is collaborative filtering [5], where past purchases and viewing decisions can be used to attract the users to purchase additional goods. Amazon, as an example uses this technology to entice the users to have additional goods. Other applications are in speech recognition [1], handwriting recognition [7], face detection [11], face recognition [22], automated driving [2], fraud detection [21], text-based sentiment analysis, email spam filtering [21], network intrusion detection, anomaly detection, classification, signal diagnosis [17], weather forecasting, anti-virus software, anti-spam software [1], genetics, image classification [14] etc. So, by using Machine Learning, the guess work can be avoided as well as time to solve a problem can be reduced thereby providing good guarantees for the solutions.

# 8    Conclusion

The techniques developed from deep learning research have already been 'impacting a wide range of signal and information processing work within the traditional and the new'. This emerging area of research in Machine Learning is expected to move it closer to Artificial Intelligence.

# References

1. Machine Learning.: http://www.mlplatform.nl/what-is-machine-learning. Accessed 01 Jan 2016
2. Mitchell, T.M.: Machine Learning, p. 421. McGraw-Hill Science/Engineering/Math (1997)
3. Deng, L., Yu, D.: Deep learning methods and applications. Found. Trends Sign. Process. 7(3–4), 197–387 (2014) [Now Publishers Inc. Hanover, MA, USA]
4. Jang, H., Yang, H.-J., Jeong, D.-S.: Object classification using CNN for video traffic detection system. In: 21st Korea-Japan Joint Workshop on Frontiers of Computer Vision (FCV), pp. 1–4. IEEE (2015)
5. Collaborative Filtering.: http://benanne.github.io/2014/08/05/spotify-cnns.html. Accessed 20 Sept 2016
6. Le, Q.V., Ngiam, J., Coates, A., Lahiri, A., Prochnow, B., Ng, A.Y.: On optimization methods for deep learning. In: Proceedings of the 28th International Conference on Machine Learning (ICML 2011), Bellevue, WA, USA, pp. 265–272 (2011)
7. LeCun, Y., Bengio, Y., Hinton, G.: Deep learning. Nature **521**, 436–444 (2015)
8. Horn, B.K.P., Schunck, B.G.: Determining optical flow. Artif. Intell., Elsevier, North Holland **17**, 185–203 (1981) [Technical Report, Artificial Intelligence Laboratory, Massachusetts Institute of Technology, Cambridge, USA (1980)]
9. Dosovitskiy, A., Fischer, P., Ilg, E., Häusser, P., Hazırbaş, C., Golkov, V., Smagt, P., Cremers, D., Brox, T.: Flownet: learning optical flow with convolutional networks. In: IEEE International Conference on Computer Vision (ICCV 2015), Santiago, Chile (2015)
10. Flying Chairs Dataset.: http://lmb.informatik.unifreiburg.de/resources/datasets/FlyingChairs.en.html. Accessed 19 Sept 2016
11. Feature Detection and Extraction.: http://in.mathworks.com/help/vision/feature-detection-andextraction.html. Accessed 14 July 2016
12. Nguyen, K., Fookes, C., Sridharan, S.: Improving deep convolutional neural networks with unsupervised feature learning. In: International Conference on Image Processing (ICIP 2015), pp. 2270–2274. IEEE (2015)

13. Chen, Y., Yang, X., Zhong, B., Pan, S., Chen, D., Zhang, H.: CNNTracker: online discriminative object tracking via deep convolutional neural network. Appl. Soft Comput. **38**, 1088–1098 (2015) [Elsevier]
14. Ouyang, W., Wang, X.: Joint deep learning for pedestrian detection, computer vision foundation. In: Proceedings of the 2013 IEEE International Conference on Computer Vision (ICCV'13), pp. 2056–2063. IEEE Computer Society, Washington, DC, USA (2013)
15. Zhang, Y., Li, X., Zhang, Z., Wu, F., Zhao, L.: Deep learning driven blockwise moving object detection with binary scene modeling. Neurocomputing **168**, 454–463 (2015) [Elsevier]. http://arxiv.org/abs/1601.07265
16. LeCun, Y., Huang, F.J., Bottou, L.: Learning methods for generic object recognition with invariance to pose and lighting. In: Proceedings of the 2004 IEEE Computer Society Conference on Computer Vision and Pattern Recognition (CVPR'04), pp. 97–104. IEEE Computer Society, Washington, DC, USA (2004)
17. Jin, L., Gao, S.: Hand-crafted features or machine learnt features? Together they improve RGB-D object recognition. In: Proceedings of the 2014 IEEE International Symposium on Multimedia (ISM'14), pp. 311–319. IEEE Computer Society, Washington, DC, USA (2014)
18. Caffe.: http://caffe.berkeleyvision.org. Accessed 30 Mar 2016
19. Deep Learning Frameworks.: https://github.com/zer0n/deepframeworks#architecture. Accessed 28 Mar 2016
20. Deep Learning Libraries.: http://machinelearningmastery.com/popular-deep-learning-libraries. Accessed 28 Mar 2016
21. Applications.: https://www.quora.com/What-are-the-practical-applications-of-deep-learning-What-are-all-the-major-areas-fields. Accessed 18 Feb 2016
22. Cascade.: http://www.svcl.ucsd.edu/projects/Cascades. Accessed 10 June 2016
23. Convnet.: http://fastml.com/object-recognition-in-images-with-cuda-convnet. Accessed 28 July 2016
24. Optical Flow.: http://www.slideshare.net/xavigiro/deep-learning-for-computer-vision-34-video-analytics-lasalle-2016. Accessed 19 Sept 2016
25. Background Subtraction.: http://www.slideshare.net/ravi5raj_88/background-subtraction. Accessed 04 Aug 2016

# A Framework for Improved Cooperative Learning Algorithms with Expertness (ICLAE)

**Deepak A. Vidhate and Parag Kulkarni**

**Abstract** A policy framework for dynamic products availability in three retailer shops in the market is investigated. Retailers can cooperate with each other and can get benefit from cooperative information by their own policies that accurately represent their goals and interests. The retailers are the learning agents in the system and use reinforcement learning to learn cooperatively from the situation. Cooperation in learning (CL) is understood in a multi-agent environment. A framework for Improved Cooperative Learning Algorithms with Expertness (ICLAE) is proposed. Expertness measuring criteria which were used in earlier work are further enhanced and improved in the proposed method. Four methods for measuring the agents' expertness are used, viz., Normal, Absolute, Positive, and Negative. The novelty of this approach lies in the implementation of the RL algorithms with expertness measuring criteria by means of Q-learning, Q($\lambda$) learning, Sarsa learning, and Sarsa($\lambda$) learning algorithms. This chapter shows the implementation results and performance comparison of these algorithms.

**Keywords** Cooperative learning · Dynamic buyer behavior · Q-learning · Reinforcement learning · Sarsa learning · Weighted strategy sharing

## 1 Introduction

In the literature of multi-agent learning, different strategies of cooperation such as joint rewards, averaging Q-tables, and Weighted Strategy Sharing are beneficial only if agents are knowledgeable and expert in relatively identical domains. It is

D.A. Vidhate (✉)
Department of Computer Engineering, College of Engineering, Pune,
Maharashtra, India
e-mail: dvidhate@yahoo.com

P. Kulkarni
iKnowlation Research Labs Pvt. Ltd., Pune, Maharashtra, India
e-mail: parag.india@gmail.com

© Springer Nature Singapore Pte Ltd. 2018                                              149
R.K. Choudhary et al. (eds.), *Advanced Computing*
*and Communication Technologies*, Advances in Intelligent Systems
and Computing 562, https://doi.org/10.1007/978-981-10-4603-2_15

assumed that they are expert and always get positive reinforcement as in the case of WSS method. This expertness measuring criteria affects the learning. So to avoid the drawback in Weighted Strategy Sharing method discussed in [1], a new expertness measuring approach, i.e., Improved Cooperative Learning Algorithms with Expertness (ICLAE) using Q-Learning, Sarsa Learning, $Q(\lambda)$ Learning, and Sarsa$(\lambda)$ Learning, is proposed and implemented here.

In a country many stores sell huge amount of goods to millions of customers. It is a good model of market chain. The sale point of each retailer confirms the information of each operation, i.e., date, customer identification code, products purchased and their amount, total amount spent, and so forth. This usually produces huge amount of data everyday. If stocked data is analyzed and turned into information then it becomes helpful. It is useful as an example to build certain forecast. It is not known that exactly which people are likely to buy this item or another item. We would not need any analysis of the data if we know it already. But because we are unknown, we can only gather information and expect to take out the answers to questions from data [2]. We may be unable to recognize the procedure totally, but still we can build a useful and good approximation. These temporary computations might not give details of everything, but may still be able to construct for some part of the data. Retailers have always encountered the difficulty of sale the right goods that would produce the highest income for them. Finding the right products for a buyer or a service is a difficult task. It is required that they must know how much the product is important for customer and what would be the future demand. In the future, retailers will suggest special package, just customized for every purchaser, soon for the instant on the whole thing (correct item to the correct purchaser at the correct period) [3]. Different parameters need to be considered in this: variation in seasons, the dependency of items, special schemes, discount, market conditions, etc. That means a correct and precise model of buyer behavior can produce attractive results. Retailers can cooperate with each other for yield maximization in different situations. Several independent tasks that can be handled by separate agents could benefit from cooperative nature of agents [3, 4].

Three retailers in the market are considered in this paper. Retailers can cooperate with each other and could benefit from cooperative nature by their own policies that accurately represent their goals and interests. The paper presents some contributions as given below:

– Three seller retail stores are considered in this paper. These stores sell a specific product and give major concessions for customers buying multiple items. Vendor's inventory strategy, refill period, and the arrival procedure of the buyer are considered to design the model. Markov decision process model has been set up for the working of the system. The actions are represented by the products purchased by the customers of different age groups in the proposed model.
– Paper shows that the retailer uses reinforcement Learning Algorithms to adapt his products effectively to achieve maximum performance.

The paper is organized as follows: Sect. 2 presents a description of proposed framework for Improved Cooperative Learning Algorithms with Expertness (ICLAE). Section 3 shows the system working, modeled by Markov decision process and setup a reinforcement learning framework for changing products in the retail market. Section 4 explains the simulation results with extended period cost as the profit benchmark and conveys the perception in the dynamic product strategy obtained through graphical result comparison. Section 5 provides the concluding remarks.

## 2 Improved Cooperative Learning Algorithms with Expertness (ICLAE)

In the proposed algorithms, each agent allocates a weight to their knowledge and makes use of it based on the amount of its teammate expertness. Figure 1 shows the representation of multi-agents for learning in cooperative scenario. Based on different expertness values, each agent allocates a weight to their knowledge and makes use of it using Q-Learning Algorithm. Every learning agent allocates some weights to other agents' Q-tables, which reflect their associated expertness and credibility [4, 5]. Then, the weighted mean of other agents' Q-tables has been taken by the learning agent and the output table is used as its fresh Q-table [5]. This method is characterized using Eq. (1):

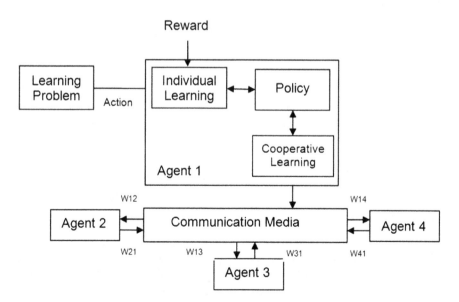

**Fig. 1** Cooperative learning framework with expertness measures

$$Q_i^{\text{new}} = \sum_{j=1}^{n} W_{ij} Q_j^{\text{old}}. \tag{1}$$

Expertness Measures: Expertness measure is performed at different levels. Expertness measures at the agent level (called as the Q-table) show the overall expertness level of an agent. The expertness measures at state level indicate how well the agent can find the optimal action in that state. Normal, absolute, positive, and negative expertness measures belong to this category. It is called as expertness measures based on learning history [5, 6]. Four methods for measuring agents' expertness are used here.

## 2.1 Expertness Measures Based on Learning History

Normal (Nrm): This measure offers more acknowledgment to those who have additional success and less failures. It is represented by a numerical addition of the reward signals as given by Eq. (2):

$$e_i^{\text{Nrm}} = \sum_{t=1}^{\text{now}} r_i(t), \tag{2}$$

where $r_i(t)$ is the amount of reinforcement signal that environment gives to agent $i$.

Absolute (Abs): It means that failures and successes, weighted by the value of reward and punishment signals, are both valuable for the agent. This is the total of the absolute cost of the reward signals given by Eq. (3):

$$e_i^{\text{Abs}} = \sum_{t=1}^{\text{now}} |r_i(t)|. \tag{3}$$

Positive (P): This criterion takes no notice of incidents that do not result in gaining rewards and considers only rewarding incidents. A total of the positive reward is used for this as given by Eq. (4):

$$e_i^P = \sum_{t=1}^{\text{pos}} r_i^+(t) \quad r_i^+(t) = 0 \quad \text{if } r_i(t) \le 0$$
$$= r_i(t) \quad \text{otherwise.} \tag{4}$$

Negative (N): It means that the agent who has more punishments knows ways are not leading to gaining rewards better and they should be considered as more expert, as the agent gets more punishments than rewards during learning. A negative is the total of the absolute cost of the negative rewards signals given by Eq. (5):

$$e_i^N = \sum_{t=1}^{now} r_i^-(t) \quad r_i^-(t) = 0 \quad \text{if} \quad r_i(t) > 0$$

$$= -r_i(t) \quad \text{otherwise.}$$

(5)

## 2.2 Cooperative Learning Algorithm with Expertness Using Q-Learning and Q(λ) Learning

Two types of learning are used, i.e., Independent Learning and Cooperative Learning Type. All of the agents are in the Independent Learning Type initially. Agent i executes $t_i$ learning based on the one-step Q-learning. Learning experiment begins at a random state and stops when the agent arrived at the target. All agents stop the Independent Learning type at the time when a particular number of Independent experiments are performed (that is called the cooperation instance). Then agents switch to Cooperative Learning Type [6]. Each learner allocates some weights to the other agents as per their expertise standards in the Cooperative Learning Type. Then, it takes a weighted mean of the others' Q-tables. Resultant Q-table is found out that is used as its new Q-table [6, 7].

**Weight-Assigning Mechanism**
To reduce the amount of communication necessary to replace Q-tables, learner makes use of the Q-tables of more skilled agents only. Incomplete weights of the less-skilled agents are taken to be zero. Learner *i* assigns the credit to the information of agent *j* as given in (6) [8]:

$$W_{ij} = \begin{cases} 1 - \alpha_i & \text{if } i = j \\ \frac{e_j - e_i}{\sum e_k - e_i} & \text{if } e_j > e_i \\ 0 & \text{otherwise} \end{cases}$$

(6)

where $0 < \alpha < 1$ is the impressibility factor and illustrate up to what extent agent *i* depends on the others knowledge. *ei* and *ej* are the expertise costs of agents *i* and *j*, correspondingly, and *n* is the total number of the agents [7, 8]. Weights assigned by each agent to other agent's information are shown in Fig. 1. w12 is the weight assigned by agent 1 to the information of agent 2 and w21 is the weight assigned by agent 2 to the information of agent 1 [9, 10]. Algorithm 1 gives the Cooperative Learning with Expertness using Q-learning and Algorithm 2 calculates the expertness of an agent based upon the estimated weights. A cooperative learning method with expertness using Q-learning and Q(λ) learning are proposed and implemented.

Algorithm 1: Cooperative Learning Algorithm with Expertness using Q learning

1. while not EndOfLearning do
    2. if InIndependentLearning Type then
    3. Independent Learning
        4. $s_i$ := FindCurrentState(); $a_i$ := SelectAction()
        5. DoAction($a_i$)
        6. $r_i$ := GetReward(); $y_i$ := GoToNextState()
        7. $v(y_i)$ := Max$_b \in$ $_{actions}$Q($y_i$,b)
        8. $Q_i^{new}$ ($s_i$,$a_i$) := (1 - $\beta_i$)$Q_i^{Old}$($s_i$,$a_i$)+ $\beta_i$($r_i$ +$\gamma_i$V($y_i$) )
        9. $e_i$ := UpdateExpertness($r_i$)
    10. else Cooperative Learning
        11. for j := 1 to n do
            12. $e_j$ := GetExpertness($A_j$)
        13. $Q_i^{new}$ := 0
        14. for j := 1 to n do
            15. $W_{ij}$ := ComputerWeights(i,j,e1......en)
            16. $Q_j^{old}$ := GetQ($A_j$)
            17. $Q_i^{new}$ := $Q_i^{new}$ + $W_{ij}$ * $Q_j^{old}$

Algorithm 2: To Calculate Expertness Value

1. GetExpertness(Aj)
2. while not EndOfLearning do
3. if NormalExpertness then
    4. $e_i^{Nrm} = \sum_{t=1}^{now} r_i(t)$
5. else if PositiveExpertness then
    6. $e_i^P = \sum_{t=1}^{now} r_i^+(t)$
7. else if NegativeExertness then
    8. $e_i^N = \sum_{t=1}^{now} r_i^-(t)$
9. else if AbsoluteExpertness then
    10. $e_i^{Abs} = \sum_{t=1}^{now} |r_i(t)|$
11. return Expertness

Algorithm 3: Cooperative Learning Algorithm with Expertness Using Sarsa Learning and Sarsa($\lambda$) Learning

## 2.3 Cooperative Learning Algorithm with Expertness Using Sarsa Learning and Sarsa($\lambda$) Learning

Algorithm 3 gives the Improved Cooperative Learning Algorithm with Expertness using Sarsa learning method. It calculates the expertness of an agent based upon the estimated weights and learns cooperatively using Sarsa learning method. Independent learning is performed based on Sarsa learning. Expertness Measures Based on Q-Table and Weight-assigning mechanism remain same as described in the earlier section for this algorithm. Cooperative learning method called with expertness using Sarsa learning and Sarsa($\lambda$) learning is implemented, based on different expertnesses, i.e., absolute, positive, and negative expertness values in the method using Sarsa and Sarsa($\lambda$) Learning Algorithm [10–12].

Algorithm 3: Cooperative Learning Algorithm with expertness using Sarsa learning

1.  while not EndOfLearning do
2.  begin
   3.  if In Independent Learning Type then
   4.  Independent Learning
      5.  $x_i :=$ FindCurrentState(); $a_i :=$ SelectAction()
      6.  DoAction($a_i$)
      7.  $r_i :=$ GetReward()
      8.  $Q_i^{new}(x_i,a_i) := (1 - \beta_i)Q_i^{Old}(x_i,a_i) + \beta_i(r_i + \gamma_i Q_i^{new}(x_i,a_i) - Q_i^{Old}(x_i,a_i))$
      9.  $x_i :=$ FindNextState(); $a_i :=$ NewAction()
      10. $e_i :=$ UpdateExpertness($r_i$)
   11. else Cooperative Learning
      12. for j := 1 to n do
      13. $e_j :=$ GetExpertness($A_j$)
      14. $Q_i^{new} := 0$
      15. for j := 1 to n do
         16. $W_{ij} :=$ ComputerWeights(i,j,e1......en)
         17. $Q_j^{old} :=$ GetQ($A_j$)
         18. $Q_i^{new} := Q_i^{new} + W_{ij} * Q_j^{old}$

# 3 Model Design

The case of wedding season as an example is considered here. Beginning from deciding the venue, booking the caterers, decoration, invitation cards, photography, beautician, cosmetics, household items, gifts, shopping of clothes, jewelry, and other accessories for bride and groom, so many activities are involved [11]. Such seasonable situations can be realistically implemented as follows: Customer who would go for clothing shop certainly will buy jewelry, footwear, and other accessories. Customer who selects the wedding venue would certainly opt for caterers and decorators. In such scenarios, retailers of different products can come together and jointly satisfy customer requirements and would achieve the benefit of an increase in the product sale [12]. Figure 2 shows a diagram for these dynamics.

Following is the mathematical description of the above model.

– Consumers enter the market by following a Poisson process flow with rate $\lambda$.
– The retailer posts per unit product price $p$ to the arriving customers.
– The retailer has finite stock facility Imax and follows a policy for refilling.

We define action set as the sale of the possible product, i.e., $A = \{$Products P1, P2, P3…P10$\}$ and action $a \in A$. So state can be described by a vector

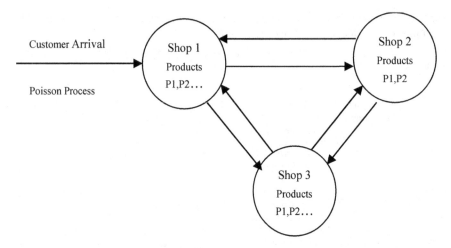

**Fig. 2** Retail store model with three retailers

X = {TID, Customer, Product Price, Month} = {TID, x1, x2, m}, TID = Transaction ID.

x1 = {Y, M, O} → customer age vector, i.e., young, middle, and Old age customer

x2 = {H, M, L} → product price vector, i.e., High, Medium, and Low

m = {1, 2, 3, 4…12} → month of product sale.

Whenever a customer places a request for a product, a decision needs to be made regarding whether to accept or deny the request. Another retailer observes the action taken by the first retailer and be prepared to sell his product. In this way as a sale in one shop increases automatically other shops get informed so they can sell their products too.

## 4 Results

First, for a given month and customer age group, the product is identified. Learning shows that for a given month and an age group which products are to be selected that are best for sale. Shop agent will understand that in a month which products are to be sold to the customers having the age group. Second, it shows that in a year, the specific number of products is purchased by the particular customer age group. Shop agent will understand that in a year, the number of products is to be sold to the customers having the different age groups. Sarsa and Sarsa($\lambda$) algorithm gives better results than Q-learning and Q($\lambda$) Learning. Result analysis is done for a year; the specific number of products is purchased by the particular customer age group. Significant improvement is seen in the results as agents learn cooperatively to increase the sale of products so as to maximize the profit.

## 4.1 Results of Improved Cooperative Learning Algorithms with Expertness

1. **Absolute Expertness**

Figure 3 shows the graph of Reward Versus Episode for Absolute Expertness for Sarsa($\lambda$) learning, Q($\lambda$) learning, Sarsa learning, and Q-learning algorithms. The number of rewards received by Sarsa learning is greater than Q-learning for each episode and the number of rewards received by Sarsa($\lambda$) learning is greater than Q ($\lambda$) learning for each episode in case of Absolute Expertness.

2. **Positive Expertness**

Figure 4 shows the graph of Reward Versus Episode for Positive Expertness for Sarsa($\lambda$) learning, Q($\lambda$) learning, Sarsa learning, and Q-Learning Algorithms. The number of rewards received by Sarsa learning is greater than Q-learning for each episode and the number of rewards received by Sarsa($\lambda$) learning is greater than Q ($\lambda$) learning for each episode in case of Positive Expertness.

3. Negative Expertness

Figure 5 shows the graph of Reward Versus Episode for Negative Expertness for Sarsa($\lambda$) learning, Q($\lambda$) learning, Sarsa learning, and Q-Learning Algorithms. The

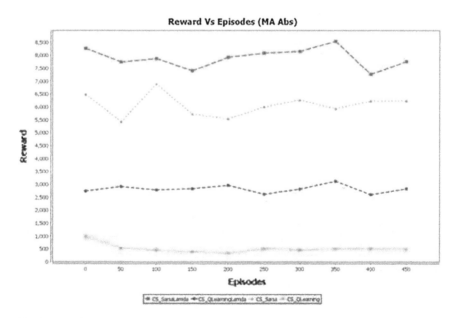

**Fig. 3** Reward versus episode for absolute expertness for four algorithms

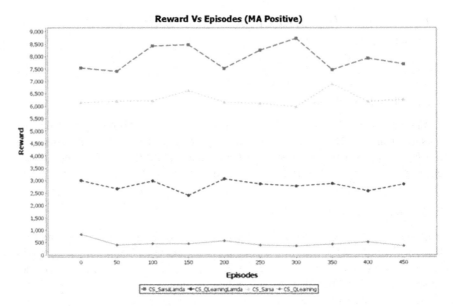

**Fig. 4** Reward versus episode for positive expertness for four algorithms

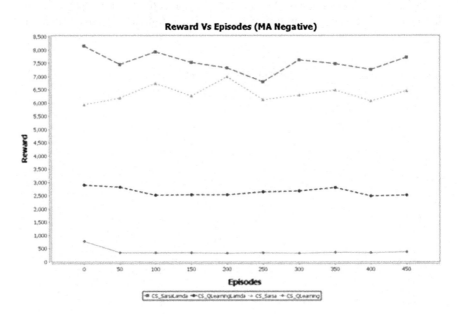

**Fig. 5** Reward versus episode for negative expertness for four algorithms

number of rewards received by Sarsa learning is greater than Q-learning for each episode and the number of rewards received by Sarsa($\lambda$) learning is greater than Q ($\lambda$) learning for each episode in case of Negative Expertness. Comparison of Reward Versus Episodes for Absolute, Positive, and Negative Expertness measure, respectively, for four algorithms is shown in Figs. 3, 4 and 5. Sarsa($\lambda$) and Sarsa learning is performing better and gives maximum reward than Q($\lambda$) and Q-learning, respectively.

# 5   Conclusion

Q-tables show the most excellent action (that is the most favorable product) for different individual states. The shop agent can compute most probable product for a known state that gives maximum profit to it by knowing the Q-function. It has demonstrated how a shop agent can successfully use reinforcement Learning Algorithms in mounting products strongly toward the profit maximization. It is believed that this is a promising approach for profit maximization in retail market setting with partial knowledge. Cooperative Learning Algorithms are more efficient and effective and produce best results. Improved Cooperative Learning Algorithm with Expertness (ICALE) shows that sharing of more knowledge and information is possible, all agents' knowledge is used equally, and problem can be jointly solved. However, this method is still unable to find more expert agents. To study and implement few communication and cooperation parameters to be considered for learning is the future scope of this paper.

# References

1. Araabi, B.N., Mastoureshgh, S., Ahmadabadi, M.N.: A study on expertise of agents and its effects on cooperative Q-learning. IEEE Trans. Evol. Comput. **14**, 23–57 (2010)
2. Vidhate, D.A., Kulkarni, P.: New approach for advanced cooperative learning algorithms using RL methods (ACLA). In: Proceedings of the Third International Symposium on Computer Vision and the Internet (Vision Net'16), pp. 12–20. ACM, DL (2016)
3. Choi, Y.-C., Student Member, Ahn, H.-S.: A survey on multi-agent reinforcement learning: coordination problems. In: IEEE/ASME International Conference on Mechatronics and Embedded Systems and Applications, pp. 81–86 (2010)
4. Abbasi, Z., Abbasi, M.A.: Reinforcement distribution in a team of cooperative Q-learning agent. In: Proceedings of the 9th ACIS International Conference on Software Engineering, Artificial Intelligence, and Parallel/Distributed Computing 978-0-7695-3263-9/08, pp. 154–160. IEEE (2008)
5. Gao, L.-M., Zeng, J., Wu, J., Li, M.: Cooperative reinforcement learning algorithm to distributed power system based on multi-agent. In: 3rd International Conference on Power Electronics Systems and Applications (Reference: K210509035) (2009)
6. Al-Khatib, A.M.: Cooperative machine learning method. World Comput. Sci. Inf. Technol. J. **1**, 380–383 (2011)

7. Vidhate, D.A., Kulkarni, P.: Enhancement in decision making with improved performance by multiagent learning algorithms. IOSR J. Comput. Eng. **1**(18), 18–25 (2016)
8. Panait, L., Luke, S.: Cooperative multi-agent learning: the state of the art. J. Auton. Agents Multi-Agent Syst. **11**, 387–434 (2005)
9. Vidhate, D.A., Kulkarni, P.: Implementation of multiagent learning algorithms for improved decision making. Int. J. Comput. Trends Technol. (IJCTT) **35**(2) (2016)
10. Tao, J.-Y., Li, D.-S.: Cooperative strategy learning in multi-agent environment with continuous state space. In: IEEE International Conference on Machine Learning (2006)
11. Dr. Berenji, H.R., Vengerov, D.: Learning, Cooperation, and Coordination in Multi-Agent Systems. Intelligent Inference Systems Corp., Technical Report, October 2000
12. Nagendra Prasad, M.V., Lesser, V.R.: Learning situation-specific coordination in cooperative multi-agent systems. J. Auton. Agents Multi-Agent Syst. **2**(2), 173–207 (1999)

# Part IV
# Advanced Computing: Data Encryption

# Bicubic Interpolation Based Audio Authentication (BIAA)

**Uttam Kr. Mondal and Jyotsna Kumar Mandal**

**Abstract** Authenticating audio songs/signals is one of the challenging issues in the domain of audio processing and analysis. Different techniques like embedding secret code in amplitude or frequency components, harmonics alternation, embedding audio/image, etc. are frequently used by the researchers for audio signal authentication. In this paper, an approach has been made to authenticate audio signal through bicubic interpolation. Bicubic interpolation is applied over songs to incorporate small changes in some specific positions and correlates with other sampled values without any major alternation in audible quality of signals. It is then molded into a suitable form within tolerable range of controlling parameters of audio quality to embed secure data/another strip of audio signal to identify it uniquely among collection of similar audio songs. A comparative study is given with similar existing techniques. Experimental results are computed based on Microsoft WAVE ("·wav") stereo sound file, and performances are analyzed which show optimal performances compared to the existing techniques.

**Keywords** Protection of intellectual property · Audio song characteristics · Bicubic interpolation · Song signal authentication · Song security

## 1 Introduction

Incorporation of highly sophisticated editing and mixing technology in music industries, chances of producing duplicate versions of original music as well as rhythm, is one of the frequent phenomena in daily life. Sometimes, it becomes

U.Kr. Mondal (✉)
Department of Computer Science, Vidyasagar University, Midnapur, West Bengal, India
e-mail: uttam_ku_82@yahoo.co.in

J.K. Mandal
Department of Computer Science & Engineering, University of Kalyani, Nadia,
West Bengal, India
e-mail: jkm.cse@gmail.com

© Springer Nature Singapore Pte Ltd. 2018                                            163
R.K. Choudhary et al. (eds.), *Advanced Computing
and Communication Technologies*, Advances in Intelligent Systems
and Computing 562, https://doi.org/10.1007/978-981-10-4603-2_16

popular than the original one. Some strong authenticating measure is needed to identify the piracy along with symmetric portions from duplicate/modified versions [1, 2]. Therefore, some controlling mechanism is needed over components of music constituents for stopping any type of alternations. Even the formation of song signal has to be managed in a way as such any minor changes may alter the overall audible quality [3, 4]. Additionally, a security/authentication code may be embedded into song signal to detect the original one from modified audio easily.

In this paper, bicubic interpolation is applied in audio signal to represent the original signal into a structural format with correlating the sampled values. A secret code is embedded for enhancing security level keeping audible quality of modified version close to the original.

The organization of the paper is as follows. Bicubic polynomial based secret key generation and fabrication of secret song are presented in Sects. 2.1 and 2.2 respectively. Extraction is given in Sect. 2.3. Experimental results are shown in Sect. 3. Conclusions are drawn in Sect. 4 that of references at the end.


## 2  The Technique

The scheme fabricates the secret key using bicubic interpolation over the fractional parts of magnitude values sampled from original song along with embedding a secret audio signal without changing the audible quality of song signal. Proposed scheme has two stages. These are bicubic polynomial based secret key fabrication and hiding authenticating strip of song as another layer authenticating code. Bicubic polynomial based secret key generation and fabrication of secret song are presented in Sects. 2.1 and 2.2, respectively.


## 2.1  Bicubic Polynomial Based Secret Key Embedding

Bicubic interpolation is one of the well-known techniques for scaling images and video, which may be opted for authenticating of audio song signal for the same. It is also observed that the bicubic interpolation provides a well-sophisticated structure of representation and generates smoother edges than that of bilinear interpolation. On the other hand, it is accomplished with other techniques such as Lagrange polynomials, cubic splines, cubic convolution algorithms, etc. [5]. The consecutive magnitude values of song signal are generally having less difference, if we take only the fraction part of magnitude values of consecutive locations of song signal and apply bicubic interpolation formula to find interpolated values from the set of fraction part of magnitude values, producing interpolated values close to fraction values which have been taken as input by the formula described in Eq. (1) as a multivariate polynomial:

$$g(x, y) = \sum_{i=0}^{3} \sum_{j=0}^{3} a_{ij} x^i y^j.$$ (1)

The values of the coefficients ($a_{ij}$) are determined with the help of the equations shown in Fig. 1, whose $p$ (in the equations of Fig. 1) is $4 \times 4$ dimensional matrix whose values are the fractional part ($p_{ij}$) of magnitude values of signal in sequence, where $1 \leq i \leq 4$ and $1 \leq j \leq 4$ [6]. The steps of the process are given in Algorithm 1.

### 2.1.1 Algorithm 1

In this technique, the original song is taken as input, and the output of the process generates the authenticated signal with embedded secret key. The details are given as follows

**Input**: Original song

**Output**: Authenticated song with embedded secret key

**Method**: At the first step of the process, frequency components of sampled values of the song are obtained by applying Fast Fourier Transform (FFT). The magnitude values of the signal (stereotype) are logically divided into $B$ blocks; applying bicubic interpolation over the blocks and modifying values of blocks, generate the authenticated audio signal. The method of song authentication with bicubic interpolation is given in the following steps.

Step 1: Find frequency components of sampled values of song signal by applying Fast Fourier Transform (FFT).

Step 2: Divide (logically) the magnitude values of the signal (stereotype) into $B$ blocks; each block consists of 16 consecutive magnitude values from a specified channel (i.e., first or second channel of stereotype song as shown in Fig. 2). The value of $B$ (number of blocks) depends on the length of song signal that can be calculated as follows: $B = $ floor ($L/N$, 0), where floor represents rounding number down, toward zero, to the nearest multiple of significance (here, significance value is 0), $L$ represents length of the song, and $N = 16$.

Step 3: Take fractional part of eight consecutive magnitude values from second half of ($N - 1$)th block of the channel 1 of stereo song and put them into first and second rows of $4 \times 4$ matrix sequentially. Similarly, take consecutive fractional parts of eight magnitude values from first half of ($N + 1$)th block of same channel (i.e., next of the previous block) and put them into third and fourth rows of the matrix sequentially (as shown in Fig. 3).

Step 4: Apply Eq. (1) to find the bicubic values of the particular portion of above matrix (step 2) and return the computed values with respective decimal part (if present) to the song signal into $N$th block (Fig. 2) sequentially.

Step 5: Repeat steps 2–4 for entire song signal alternately for two channels.

Step 6: Apply inverse FFT to regenerate the sampled values of modified signal.

| Coefficient $a_{ij}$ | Value |
|---|---|
| $a_{00}$ | $P_{11}$ |
| $a_{01}$ | $-\frac{1}{2}P_{10} + \frac{1}{2}P_{12}$ |
| $a_{02}$ | $P_{10} - \frac{5}{2}P_{11} + 2P_{12} - \frac{1}{2}P_{13}$ |
| $a_{03}$ | $-\frac{1}{2}P_{10} + \frac{3}{2}P_{11} - \frac{3}{2}P_{12} + \frac{1}{2}P_{13}$ |
| $a_{10}$ | $-\frac{1}{2}P_{01} + \frac{1}{2}P_{21}$ |
| $a_{11}$ | $\frac{1}{4}P_{00} - \frac{1}{4}P_{02} - \frac{1}{4}P_{20} + \frac{1}{4}P_{22}$ |
| $a_{12}$ | $-\frac{1}{2}P_{00} + \frac{5}{4}P_{01} - P_{02} + \frac{1}{4}P_{03} + \frac{1}{2}P_{20} - \frac{5}{4}P_{21} + P_{22} - \frac{1}{4}P_{23}$ |
| $a_{13}$ | $\frac{1}{4}P_{00} - \frac{3}{4}P_{01} + \frac{3}{4}P_{02} - \frac{1}{4}P_{03} - \frac{1}{4}P_{20} + \frac{3}{4}P_{21} - \frac{3}{4}P_{22} + \frac{1}{4}P_{23}$ |
| $a_{20}$ | $P_{01} - \frac{5}{2}P_{11} + 2P_{21} - \frac{1}{2}P_{31}$ |
| $a_{21}$ | $-\frac{1}{2}P_{00} + \frac{1}{2}P_{02} + \frac{5}{4}P_{10} - \frac{5}{4}P_{12} - P_{20} + P_{22} + \frac{1}{4}P_{30} - \frac{1}{4}P_{32}$ |
| $a_{22}$ | $P_{00} - \frac{5}{2}P_{01} + 2P_{02} - \frac{1}{2}P_{03} - \frac{5}{2}P_{10} + \frac{25}{4}P_{11} - 5P_{12} + \frac{5}{4}P_{13} + 2P_{20} - 5P_{21} + 4P_{22} - P_{23} - \frac{1}{2}P_{30} + \frac{5}{4}P_{31} - P_{32} + \frac{1}{4}P_{33}$ |
| $a_{23}$ | $-\frac{1}{2}P_{00} + \frac{3}{2}P_{01} - \frac{3}{2}P_{02} + \frac{1}{2}P_{03} + \frac{5}{4}P_{10} - \frac{15}{4}P_{11} + \frac{15}{4}P_{12} - \frac{5}{4}P_{13} - P_{20} + 3P_{21} - 3P_{22} + P_{23} + \frac{1}{4}P_{30} - \frac{3}{4}P_{31} + \frac{3}{4}P_{32} - \frac{1}{4}P_{33}$ |
| $a_{30}$ | $-\frac{1}{2}P_{01} + \frac{3}{2}P_{11} - \frac{3}{2}P_{21} + \frac{1}{2}P_{31}$ |
| $a_{31}$ | $\frac{1}{4}P_{00} - \frac{1}{4}P_{02} - \frac{3}{4}P_{10} + \frac{3}{4}P_{12} + \frac{3}{4}P_{20} - \frac{3}{4}P_{22} - \frac{1}{4}P_{30} + \frac{1}{4}P_{32}$ |
| $a_{32}$ | $-\frac{1}{2}P_{00} + \frac{5}{4}P_{01} - P_{02} + \frac{1}{4}P_{03} + \frac{3}{2}P_{10} - \frac{15}{4}P_{11} + 3P_{12} - \frac{3}{4}P_{13} - \frac{3}{2}P_{20} + \frac{15}{4}P_{21} - 3P_{22} + \frac{3}{4}P_{23} + \frac{1}{2}P_{30} - \frac{5}{4}P_{31} + P_{32} - \frac{1}{4}P_{33}$ |
| $a_{33}$ | $\frac{1}{4}P_{00} - \frac{3}{4}P_{01} + \frac{3}{4}P_{02} - \frac{1}{4}P_{03} - \frac{3}{4}P_{10} + \frac{9}{4}P_{11} - \frac{9}{4}P_{12} + \frac{3}{4}P_{13} + \frac{3}{4}P_{20} - \frac{9}{4}P_{21} + \frac{9}{4}P_{22} - \frac{3}{4}P_{23} - \frac{1}{4}P_{30} + \frac{3}{4}P_{31} - \frac{3}{4}P_{32} + \frac{1}{4}P_{33}$ |

**Fig. 1** Coefficients of bicubic interpolation

As consecutive values of song signal are not varying rapidly [1], therefore, replacing fractional part of a block magnitude values by the resultant bicubic interpolated values of previous and next blocks will not change its audible quality as well. In the case of monotype song, the above process is applied for magnitude values of a single channel. All substituted bicubic interpolated values will maintain a relationship for uniquely identifying the original signal.

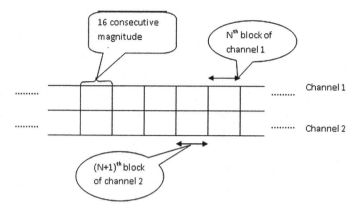

**Fig. 2** Magnitude values of sterotype song grouped into blocks

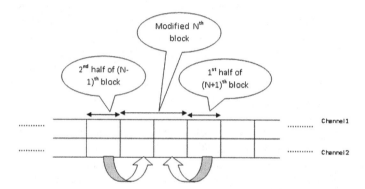

**Fig. 3** $N$th block as a result of bicubic interpolation of half of $(N-1)$th and $(N + 1)$th blocks

## 2.2 Fabrication of Secret Code

A secret hidden code has been embedded into the signal for providing another level of security to identify it uniquely from collection of similar songs. A strip of song or a selected part of the same song may be embedded with the song signal.

If we add the amplitude values of hidden song directly with any or both channels of original song, the effect over audible quality would be easily detected only by hearing the song. Therefore, the intensity of hidden song signal's amplitude values has to be reduced without any information lost. Two factors are to be maintained while embedding it with the song signal. These are the length of hidden song and the intensity of hidden song signal's amplitude values. The hidden song will be embedded in the positions of bicubic interpolated values where its length will be equal to half of total number of bicubic interpolated blocks $\times 16$. In case of stereotype hidden song, its length again reduced by half as embedding is done in

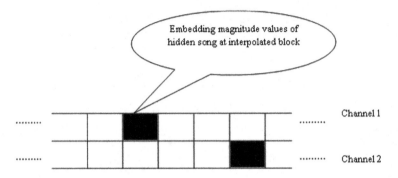

**Fig. 4** Embedding hidden song (from Fig. 3)

two separate channels. To reduce the intensity of amplitude values of hidden song, following measures are incorporated in the technique. Separate each digit from a particular amplitude value and represent it by lower fractional number (1/10000). For example, let the amplitude value is 0.6802, individual digits are 6, 8, 0, and 2. Represent it by lower fractional values as 0.0006, 0.0008, 0, and 0.0002. An additional lower fractional value may be added for representing positive (+) or negative (−) amplitude value like 0.0005 for positive and 0.0007 for negative, i.e., each amplitude value would be represented by the combination of 5 fractional values. The amplitude value is taken whereas precision level is 4 [using Matlab 6.5 for a WAVE (".wav") stereo sound file on Microsoft Windows XP environment]. Selected length of hidden song signal again to be reduced by 5 as each magnitude value is represented by a combination of 5 lower magnitude values and the set of lower magnitude values will use for embedding with the original structured song (as described in Sect. 2.1).

The amplitude values are embedded with the bicubic interpolated values of signal as illustrated in Fig. 4. Let, $O\_S[c, i]$ is the lower magnitude value of original song signal for channel c and $H\_S[i]$ is the divided amplitude value hidden song for ith position. The process of embedding hidden song signal is given in Algorithm 2.

### 2.2.1   Algorithm 2

**Input**: Authenticated song signal and secret key (using bicubic interpolation).
**Output**: Authenticated song with embedded secret audio.
**Method**: The method of embedding secret song is given in following steps.

Step 1: Add the lower amplitude values of hidden song in the consecutive positions of first bicubic interpolated block as represented by expression (2):

$$OS[c, i] = OS[c, i] + HS[i], \tag{2}$$

where $i$ represents the $i$th magnitude value of first bicubic interpolated block for $c$th (=1 or 2) channel. Repeat the process of embedding of hidden song's amplitude values in the entire bicubic interpolated block for consecutively 16 positions (Fig. 4).

Step 2: Same set of amplitude values are also to be added with next bicubic interpolated block in the alternate channel for determining any alternation of song signal under any type of intentional or unintentional attacks.

Step 3: Repeat the process of embedding of hidden song's amplitude values for the entire song.

Though the embedded amplitude values (of hidden song) are very small, therefore, appending extra values with original song (cover song) will not alter its overall audible quality. Therefore, comparing the embedded part of two channels with the result of bicubic interpolation of fraction parts of previous and next blocks of corresponding channel, the hidden song can be regenerated during extraction.

## 2.3 Extraction

The extraction is done using similar computations as embedding technique. Extracting the transformed values of bicubic interpolated values, the original signal can be perfectly regenerated. The decoding algorithm is given in Algorithm 3 in Sect. 2.3.1.

### 2.3.1 Algorithm 2

During decoding, the authenticated song is taken as input, and the output of the process is the original song extracting secret signal. The details are given as follows:

**Input**: Song with embedded authenticating code.

**Output**: Decoded signal.

**Method:** Regeneration of original signal through extraction of embedded information/signal.

Step 1: Apply technique as described in Sect. 2.1 to find the bicubic interpolation values of embedded song for the specific channel.

Step 2: Subtract the computed value from modified magnitude value to obtain amplitude values of secret signal for a specific position.

Step 3: Continue step 3 until total amplitude values of secret song are completely restored.

## 3 Experimental Results

Applying encoding and decoding techniques over original and modified songs for duration of 10 s and the associated results are discussed in two sections. Section 3.1 represents results associated with BIAA and Sect. 3.2 shows a comparison with similar existing techniques. Section 3.2 is used to determine the estimated error in the original song Fig. 5.

### 3.1 Results

Finding experimental results, a strip of 10 sections song is selected and corresponding results are observed. The chosen classical song is '100 Miles From Memphis', sung by Sheryl Crow. Table 1 shows the sampled values of the song which is a two-dimensional matrix. The amplitude–time graph of the input signal is represented in Fig. 5. BIAA is applied on this signal and the generated output of this process is given in Fig. 6. The ratio signal of frequencies of original and secret code embedded songs is shown in Fig. 7. As a result, the audible quality of modified song is not altered as the deviation between original and modified signals is minimum (as shown in Fig. 6).

Table 1 shows the values of sampled data array $x(n, 2)$ of the original song and graphical representation of the same is shown in Fig. 5 considering all rows (451584) of $x(n, 2)$ respectively. Figure 6 represents the authenticated signal, whereas the different magnitude values of original and authenticated signals are shown in Fig. 7. The authenticated song is shown in Fig. 7.

| Table 1 Sampled data array $X(n, 2)$ | S. No. | $x(k, 1)$ | $X(k, 2)$ |
|---|---|---|---|
| | ... | ... | ... |
| | | 0 | 0.0001 |
| | | 0.0000 | 0.0000 |
| | | −0.0009 | −0.0009 |
| | | −0.0006 | −0.0007 |
| | | −0.0012 | −0.0012 |
| | | −0.0014 | −0.0014 |
| | | −0.0016 | −0.0017 |
| | | −0.0023 | −0.0022 |
| | | −0.0027 | −0.0027 |
| | | −0.0022 | −0.0021 |
| | ... | ... | ... |

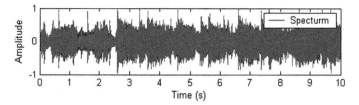

**Fig. 5** Original song ('100 Miles from Memphis' sang by Sheryl Crow)

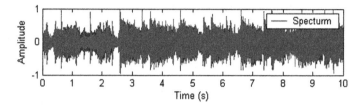

**Fig. 6** Modified song with secure code

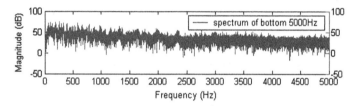

**Fig. 7** The difference of magnitude values between signals shown in Figs. 4 and 1

## 3.2 Comparisons

Algorithms [7], those available in the literature for embedding information into audio signals, usually do not care much about the audible quality. In the present technique, a special care has been taken for authenticating audio with minimum change in the quality of song. A comparative study of properties of the present method with an existing technique (Data hiding through phase manipulation of audio signals (DHPMA)) [8] over the input signal (16 bit stereotype, sampled rate 44.1 kHz) is represented in Tables 2, 4, and 5. From these tables, it is shown that a lower average absolute difference (AD) is achieved which signifies lesser error in authenticated song.

Average absolute difference (AD) is used as the dissimilarity measurement between original song and authenticated song, whereas a lower value of AD signifies lesser error in the authenticated song. Normalized average absolute difference (NAD) is used for measuring quantization error (ranging between 0 and 1) whereas

**Table 2** MD, AD, NAD, MSE and NMSE

| S. No. | Statistical parameters for differential distortions | Value (using BIAA) | Value (using DHPMA) |
|--------|------------------------------------------------------|---------------------|----------------------|
| 1 | MD | 0.0117 | 3.6621e−004 |
| 2 | AD | 0.0012 | 2.0886e−005 |
| 3 | NAD | 0.0103 | 0.0063 |
| 4 | MSE | 2.2975e−006 | 1.4671e−009 |
| 5 | NMSE | 1.0155e+004 | 8.4137e−005 |

**Table 3** SNR, PSNR, BER and MOS

| Audio (10 s) | SNR | PSNR | BER | MOS |
|--------------|---------|---------|-----|-----|
| Song1 | 40.0668 | 56.2154 | 0 | 5 |
| Song2 | 33.5932 | 50.5537 | 0 | 5 |
| Song3 | 32.2014 | 46.1211 | 0 | 5 |
| Song4 | 35.2814 | 52.9682 | 0 | 5 |
| Song5 | 40.1995 | 55.4529 | 0 | 5 |

**Table 4** SNR and PSNR

| S. No. | Statistical parameters for differential distortions | Value (using BIAA) | Value (using DHPMA) |
|--------|------------------------------------------------------|---------------------|----------------------|
| 1 | Signal-to-noise ratio (SNR) | 40.0668 | 40.7501 |
| 2 | Peak signal-to-noise ratio (PSNR) | 56.2154 | 45.4226 |

**Table 5** NC and QC

| S. No | Statistical parameters for correlation distortions | Value (using BIAA) | Value (using DHPMA) |
|-------|----------------------------------------------------|---------------------|----------------------|
| 1 | Normalized cross-correlation (NC) | 1 | 1 |
| 2 | Correlation quality (QC) | −0.2238 | −0.5038 |

mean square error (MSE) represents the cumulative squared error between original and modified signals. Therefore, a lower value of MSE is expected for an embedding technique. The higher the PSNR (Peak signal-to-noise ratio) and lower SNR (Signal-to-noise ratio) assured the good quality of embedded signal. Thus, the benchmarking parameters (NAD, MSE, NMSE, SNR, and PSNR) in proposed method show better performance with minimum change in audible quality. The values of SNR and PSNR are shown in Table 4, whereas the normalized cross-correlation (NC) and correlation quality (QC) of the present method with DHPMA are represented in Table 5. PSNR, SNR, BER (Bit Error Rate), and MOS (Mean opinion score) values are given in Table 3. Figure 8 describes the overall results of the experimental test. The quality rating (Mean opinion score) is computed by using (3):

**Fig. 8** Performance for different audio signals

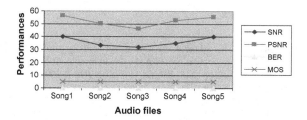

**Table 6** Quality rating scale

| Rating | Impairment | Quality |
|--------|------------|---------|
| 5 | Imperceptible | Excellent |
| 4 | Perceptible, not annoying | Good |
| 3 | Slightly annoying | Fair |
| 2 | Annoying | Poor |
| 1 | Very annoying | Bad |

$$\text{Quality} = \frac{5}{1 + N * \text{SNR}}, \tag{3}$$

where $N$ represents normalization constant. The ITU-R Rec. 500 quality rating is used for quality rating on a scale from 1 to 5 [9] and represented in Table 6.

## 4 Conclusion and Future Work

In this paper, an algorithm for encoding the fraction part of magnitude values with the help of cubic interpolation method in selective part of signal has been proposed. On the other hand, considering only fraction part of magnitude values of consecutive locations of song signal and apply bicubic interpolation produces interpolated values close to fraction values which give better results compared to other algorithms (with minimum humming distance).

The third layer of security is incorporated considering the extra deducted values of fractional part of magnitude values by sequentially adding in lower/higher frequency region (less than 20 Hz and greater than 20 kHz) which will not alter the song quality but ensure to identify the distorted properties of signal. Additionally, the proposed technique is very easy to implement.

The technique can also be extended for embedding image in place of audio. The allowable volume of data that suited for embedding with audio will be estimated in future with considering all possibilities.

# References

1. Mondal U.K., Mandal, J.K.: A novel technique to protect piracy of quality songs through amplitude manipulation (PPAM). In: International Symposium on Electronic System Design (ISED 2010), pp. 246–250 (2010). ISBN 978-0-7695-4294-2
2. Erten, G., Salam, F.: Voice output extraction by signal separation. In: ISCAS '98, vol. 3, pp. 5–8 (1998). ISBN 07803-4455-3
3. Blackledge, J.M., Al-Rawi, A.I., Hickson, R.: Multi-channel audio information hiding. In: Proceedings of the 15th International Conference on Digital Audio Effects (DAFx-12), York, UK, 17–21 Sept 2012
4. Zhang, J.: Audio dual watermarking scheme for copyright protection and content authentication. Int. J. Speech Technol. **18**(3), 443–448 (2015). ISSN 1381-2416
5. Keys, R.G.: Cubic convolution interpolation for digital image processing. IEEE Trans. Acoust. Speech Signal Process. **ASSP-29**(6), 1153–1160 (1981)
6. http://en.wikipedia.org/wiki/Bicubic_interpolation, last accessed January 11, 2013 at 11:05 PM IST
7. Katzenbeisser, S., Petitcolas, F.A.P.: Information Hiding Techniques for Steganography and Digital Watermarking. Artech House, Norwood, MA (2000). ISBN 978-1-58053-035-4
8. Dong, X., Bocko, M.F., Ignjatovic, Z.: Data hiding via phase manipulation of audio signals. In: IEEE International Conference on Acoustics, Speech, and Signal Processing (ICASSP 2004), vol. 5, pp. 377–380 (2004). ISBN 0-7803-8484-9
9. Arnold, M.: Audio watermarking: features, applications and algorithms. In: IEEE International Conference on Multimedia and Expo, vol. 2, pp. 1013–1016, New York, NY (2000)

# Hierarchical Metadata-Based Secure Data Retrieval Technique for Healthcare Application

**Sayantani Saha, Priyanka Saha and Sarmistha Neogy**

**Abstract** The metadata representation in the context of data privacy is one of the biggest challenges in data security. The segregation of the data attributes for secure data retrieval as well as data storage is the primary motivation for this work. The metadata-based data retrieval is performed for providing a layer of protection on the actual data set and also for efficiently searching the location of encrypted data. The paper aims to provide a technique for representing the metadata of the encrypted e-health data for a secure data retrieval process. The hierarchical representation of metadata helps in retrieving the data in an efficient way so that access to the sensitive information could be controlled. The paper proposes a novel technique for metadata design for secure data retrieval. E-health data is fragmented over multiple servers based on sensitive attribute and sensitive association. A brief overview of the data protection and data retrieval techniques with respect to the proposed metadata representation is also presented.

**Keywords** Metadata · Fragmentation · Indexing · Optimum encryption · Sensitive data · Health data

## 1 Introduction

According to the 6th Annual IDC Digital Universe study [1], the digital universe will grow by a factor of 300 from 2005 to 2020, i.e., from 130 to 40,000 exabytes, and Embedded and Medical Information in the cloud in 2020 would constitute

S. Saha (✉)
School of Mobile Computing and Communication, Jadavpur University, Kolkata, India
e-mail: sayantanircc@gmail.com

P. Saha · S. Neogy
Department of Computer Science and Engineering, Jadavpur University, Kolkata, India
e-mail: priyanka.saha@gmail.com

S. Neogy
e-mail: sarmisthaneogy@gmail.com

© Springer Nature Singapore Pte Ltd. 2018
R.K. Choudhary et al. (eds.), *Advanced Computing and Communication Technologies*, Advances in Intelligent Systems and Computing 562, https://doi.org/10.1007/978-981-10-4603-2_17

175

8.5% of the total data. Rapid connectivity and huge infrastructure of organization help in creating data exponentially and data is being stored by database provider. A service provider may be honest but curious and at times, even malicious. So protection of data from service providers as well as from any intruder is needed. There are some preventive measures [2], e.g., which can be applied toward protecting data. Access control is a mechanism which regulates the entry of any user into the system and also controls the read and write access for any particular data. Inference control is used for preventing intruders to infer sensitive information from database. Finally, data encryption is used for protecting sensitive or confidential data in the storage or during transmission. As data resides in database for a large time during its life cycle, protection of data at storage phase is of utmost importance. Effective, efficient, and cost-protective data storage is a challenge that needs to be fulfilled. So metadata-based storage for efficient retrieval of data from data store is proposed here. Metadata keeps track of data location.

An advantage of using metadata-based storage is that it helps in fast query processing, privacy of query computation, and easy and efficient data retrieval. It also helps in maintaining sensitive data and association among them by hiding actual data. Metadata-based storage with proper data fragmentation technique provides wholesome protection of data where fragments can be concatenated only by authorized user.

The rest of the paper is organized as follows. Section 2 discusses related works. Section 3 presents the architecture of proposed model and the data retrieval process. Section 4 discusses the results, and analyzes and compares the efficiency of data retrieval between a metadata-based model and non-metadata-based model. We conclude our work in Sect. 5.

## 2 Related Work

For the need of privacy, confidentiality and integrity of sensitive data, different works [3] have been carried out in different areas of data protection like in access control, query execution on encrypted data, metadata concept for data retrieval, etc.

Authors in [4] proposed a privacy preserving data querying technique where SQL queries are evaluated on encrypted data hosted by Database Service Providers. Wang and Lakshmanan [5] have discussed the techniques to prevent any attacker from learning any sensitive information and thus the attacker is unable to improve his belief properties after watching series of queries and their answers. In [6, 7] the authors have stressed on the sensitivity association rather than the sensitive attributes. The authors in [8] presented metadata-based database design technique. A user query dynamically generates referential integrity and dropped the details after query execution. However, they have applied this technique over oracle 11g database and rely on Oracle internal TDE system. The authors in [9] also presented a data confidentiality technique by combining fragmentation and encryption.

They used Dynamic Hashing Fragmented Components to fragment the cloud dataset and encrypt the sensitive data.

Therefore, keeping in mind different issues like scalability, performance, recovery with secure data storage, and retrieval, we present here hierarchical metadata representation that will balance the visibility and confidentiality of the e-health data.

## 3   Proposed Technique

We have considered a remote healthcare application where health information regarding the patients are available on demand basis to authorized users like doctor, nurse or any administrative professional, etc. Here we consider that chunks of information are available in fragments and access to different fragments (single/multiple) is defined in the authorization policy. In our work, we have designed such hierarchy of sensitive information for secure storage and access. Sensitive attributes are classified and then fragmented, indexed, encrypted for fine-grained data access.

In our healthcare application let us take the original Relation $R$ of the system:

$R = \{$patient_name, guardian_name, Date_of_Birth, gender, area, phno,

$\qquad$ SSN, PID, policyno, systolic_rate, diastolic_rate, pulse_rate,

$\qquad$ hemoglobin_level, blood_group, disease_name, disease_details, date$\}$.

### 3.1   Fragmentation

**Attribute Sensitivity and Confidentiality Constraints** (**CC**): Attributes like patient id, name, disease details, etc. that may reveal sensitive information of an individual are considered as sensitive. Often it is observed that rather than the value of any attribute, associations among the values of attributes are more sensitive and can reveal confidential information. Confidentiality Constraints [10] captures all security requirements as a set of attributes of original relation $R$ whose joint occurrence and visibility need to be protected.

From the above relation $R$, Confidentiality Constraints can be defined as follows:

$C_0 = \{$SSN$\}$

$C_1 = \{$PATIENT_NAME, DISEASE_NAME, DISEASE_DETAILS$\}$

$C_2 = \{$PATIENT_NAME, POLICY$\}$

$C_3 = \{$DISEASE_NAME, DISEASE_DETAILS, JOB$\}$

$C_4 = \{$PATIENT_NAME, JOB$\}$

**Hierarchical Classification**: Data confidentiality can be obtained by proper classification of the sensitive association. Following technique is used for hierarchical classification of the metadata.

A benchmark set of queries and results in a typical healthcare application have been considered here. This data is actually used to rank different attributes used as reference index to fetch different query data. The higher the rank index value, the higher the chance of revealing the correlated or associated data. For example if patient_id is used to refer to patient general and health information and disease details then if the patient_id is revealed corresponding information may be revealed to some unauthorized entity. Therefore, we need to design the metadata in such a way that patient_id will not be the referring index of all these fragments. In our scenario for health data and health data queries, we have two sets of data fragmentation factors: (1) Confidentiality Constrained Factor (CCVattr) and (2) Rank Index Value (RIV). Now, applying Eq. (1) we calculate the sensitivity factor of particular attributes and its association and generate hierarchical classification of metadata:

$$\text{Sensitivity Factor} = C_{err} \times \frac{\text{CCVattr}_i + \text{RIV}_i}{\sum_{i=0}^{n} \text{RIV}}, \tag{1}$$

where $C_{err}$ is the calculation error for classification. The attributes with high sensitivity factor will be grouped together and that group is referred as high sensitive. The attributes with low sensitivity factors are grouped together and referred as low sensitive. With the benchmark data, we evaluated 50 query sets and calculated sensitivity factors of different attributes and identified high sensitive attributes like Patient_id = 0.83, Disease_name = 0.66, etc. High sensitive attributes and its associated value will be fragmented at a different hierarchy. One common attribute is present in all fragments that are used for relating the fragments, of course, on basis of authorization. From the relation $R$ of the system, following fragments are created while maintaining the requirements in Confidentiality Constraints.

High sensitive
S1 = {PATIENT_ID, TID}
S2 = {DISEASE_ID, DISEASE_NAME}{DISEASE_ID, TID}
Low sensitive
F1 = {TID, AREA, PATIENT_NAME, GUARDIAN_NAME, DOB, GENDER, PHNO, $(\text{SSN}^k)$}
F2 = {TID, $(\text{POLICY}^k)$, JOB, AREA}
F3 = {TID, SYSTOLIC_RATE, DIASTOLIC_RATE, PULSE_RATE, BLOOD_GROUP,
HEMOGLOBIN_LEVEL, VISITING_DATE}
F4 = {TID, $(\text{DISEASE\_NAME}^k)$, $(\text{DISEASE\_DETAILS}^k)$}.

Figure 1 shows how fragments are stored in different servers. When data retrieval is performed data is fetched from the fragments.

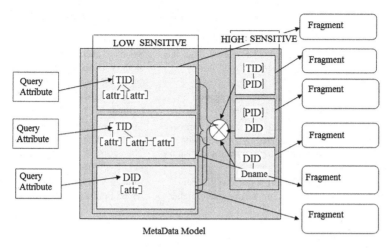

**Fig. 1** Hierarchical metadata and fragments stored in different servers

**Table 1** Patient relation

| TID | Patient_name | Job | Disease | DOB |
|-----|--------------|-----|---------|-----|
| 103 | Ishita | Engineer | Renal disease | 1984 |
| 104 | Rohan | Engineer | Gastritis | 1978 |
| Indexed version of the relation is given below | | | | |
| 103 | ψ | μ | δ | Ω |
| 104 | ¥ | μ | ω | π |

## 3.2 Indexing

Indexing helps in query evaluation on encrypted data. Different indexing techniques [10] are used for different conditions as shown in Table 1.

## 3.3 Encryption

For confidentiality purpose, queries need to be evaluated on encrypted data. The metadata information will help fetch the correct data for the user.

A double layer encryption [11] is used to encrypt the detailed data. Based on the sensitivity the data segments are encrypted first using first set of keys, i.e., general data encrypted with Key1, low sensitive data encrypted with Key2, and high sensitive data encrypted with Key3. Then another layer of encryption, using database key, is applied over the data when the data resides at database.

**Table 2**  Comparison of secure data retrieval approach between proposed and [8]

| Aspects | Trivedi, Zavarsky and Butakov [8] | Proposed |
|---|---|---|
| Data privacy | Oracle inbuilt TDE is used. This data base administrator could infer information | Selective encryption technique along with another layer of encryption is used |
| Data fragmentation | No clear Attribute segregation technique is described | A novel attribute segregation technique is proposed here |
| Query execution | Disease-related information can be easily perceived by any user who has access to patient generic information | There are different user access levels to different data attributes. The access is very much restricted in our approach |
| Information Retrieval | Data location is not specified so searching time will be much higher | Efficient indexing technique is used which optimized data location |

## 3.4  Query Evaluation

Metadata contains information about a patient's residing area, tid which will be used for referring and joining the fragments. In our proposed kiosk-based model, direct indexing technique has been used for hiding the actual location of the patient. From the metadata table user gets TID and AREAID for a patient.

Original Query will look like:

Select disease_details from table_T where patient_name="priyanka saha" and date="21/05/2016";

Now the query will be evaluated as follows:

Select area from direct_index_table where areaid='K';

In our proposed technique, query will be translated as follows:

- Select TID from metadata_table1 where patient_name = priyanka saha" and area = "area";
- Select PID from metadata_table2 where TID="TID";
- Select DID from metadata_table3 where PID="PID";
- Select * from metadata_table4 where DID="DID" and TID = "TID" sort by date.

Now the query execution requires multiple join operations of high sensitive attributes. According to the defined access control policy only authorized user can join and get details and the corresponding decryption key for decryption of the information.

## 4  Result and Analysis

With the help of our experimental setup designed over database Cassandra [12], the performance of the proposed technique and execution time of the queries are tested. We calculated data retrieval time for different types of queries—simple and

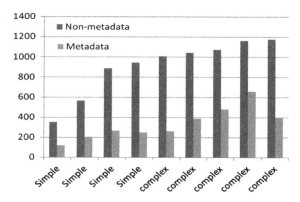

**Fig. 2** Time comparison for data retrieval from one kiosk: metadata-based model and non-metadata-based model

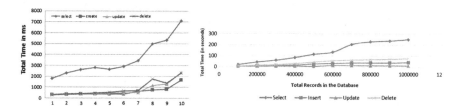

**Fig. 3** Performance comparison for data retrieval of proposed technique with [8]

complex. Complex queries refer to queries with multiple join operations. Figure 2 shows search time analysis for both metadata-based search and non-metadata-based search. Proposed technique shows considerable less time in searching the data.

It is apparent from Fig. 3 that for same input details, non-metadata-based data retrieval takes more time than metadata-based data retrieval. Table 2 shows that our technique performs considerably better than proposed by authors in [8] in terms of security and efficiency. The reasons are database (Cassandra over Oracle) efficiency and data retrieval strategy. Linear searching and joining operations are used in hierarchical metadata-based approach. Hence the complexity of our approach is $O(n)$.

## 5  Conclusion

In the proposed model, security in data storage and secure data retrieval are considered. A metadata-based model is proposed for enabling efficient query evaluation while maintaining secure storage, privacy and confidentiality of data. The primary aim is to provide secure data management. It is necessary to identify what information needs to be revealed to users for carrying out their tasks while hiding away as much excess information as possible that may potentially disclose sensitive information. The results are analyzed with a comprehensive set of experiments. The results show that the proposed techniques can deliver efficient query evaluation while ensuring enforcement of the security constraints and maintaining confidentiality of sensitive data.

**Acknowledgements** This work is supported by Information Technology Research Academy (ITRA), Government of India, ITRA-Mobile grant ITRA/15(59)/Mobile/RemoteHealth/01.

# References

1. http://www.computerworld.com/article/2509588/data-center/world-s-data-will-grow-by-50x-in-next-decade–idc-study-predicts.html, https://www.emc.com/collateral/analyst-reports/idc-the-digital-universe-in-2020.pdf
2. Hasan, R., Myagmar, S., Lee, A.J., Yurcik, W.: Toward a threat model for storage systems. In: Proceedings of the 2005 ACM Workshop on Storage Security and Survivability Storage SS'05, pp. 94–102. ACM New York (2005)
3. Weng, L., Amsaleg, L., Morton, A., Marchand-Maillet, S.: A privacy-preserving framework for large-scale content-based information retrieval. IEEE Trans. Inf. Forensics Secur. **10**(1), 152–167 (2015)
4. Hacigümüş, H., Iyer, B., Li, C., Mehrotra, S.: Executing SQL over encrypted data in the database-service-provider model. In Proceedings of the 2002 ACM SIGMOD International Conference on Management of data, pp. 216–227. ACM, NY (2002). http://www.ics.uci.edu/~chenli/pub/sigmod02.pdf
5. Wang, H., Lakshmanan, L.V.: Efficient secure query evaluation over encrypted XML databases. In: Proceedings of the 32nd International Conference on Very large Data Bases. pp. 127–138. VLDB Endowment (2006). http://www.vldb.org/conf/2006/p127-wang.pdf
6. di Vimercati, S.D.C., Erbacher, R.F., Foresti, S., Jajodia, S., Livraga, G., Samarati, P.: Encryption and fragmentation for data confidentiality in the cloud. In: Aldini, A., et.al. (eds.) Foundations of Security Analysis and Design, FOASD VII LNCS 8604, pp. 212–243. Springer International Publishing, Switzerland (2014)
7. di Vimercati, S.D.C., Foresti, S., Samarati, P.: Selective and fine-grained access to data in the cloud. In: Jajodia S., et al. (eds.) Secure Cloud Computing, pp. 123–148. Springer Science + Business Media, New York (2013). http://spdp.di.unimi.it/papers/dfs-aro2013.pdf
8. Trivedi, D., Zavarsky, P., and Butakov, S.: Enhancing relational database security by metadata segregation. Procedia Comput. Sci. **94**, 453–458 (2016) (Shakshuki, E. (ed.) Elsevier B.V., Amsterdam)
9. Jegadeeswari, S., Dinadayalan, P., Gnanambigai, N.: A Neural data security model: ensure high confidentiality and security in cloud data storage environment. In: Lloret Mauri J., et al. (eds.) International Conference on Advances in Computing, Communications and Informatics (ICACCI, Kochi, India), pp. 400–406. IEEE (2015)
10. De Capitani di Vimercati, S., et al.: Private data indexes for selective access to outsourced data. In: Proceedings of the 10th annual ACM workshop on Privacy in the Electronic society, ACM NY(2011)
11. Saha, S., Das, R., Datta, S. Neogy, S.: A cloud security framework for a data centric WSN application: In Proceedings of the 17th International Conference on Distributed Computing and Networking, pp. 39. ACM (2016)
12. Cassandra: http://cassandra.apache.org/

# Intelligent Multiple Watermarking Schemes for the Authentication and Tamper Recovery of Information in Document Image

**K.R. Chetan and S. Nirmala**

**Abstract** Existing works on document image watermarking provide same level of protection for information present in the source document image. However, in a document image the distribution of the information contents influences on the level of protection required. This necessitates application of multiple watermarking techniques on the source document image. In this paper, novel intelligent multiple watermarking techniques are proposed. The source document image is divided into blocks of the same dimension. For each block, appropriate type of watermarking is decided based on the type of block which is determined automatically using gradient binarized technique. The blocks with regeneratable information are protected using semi-fragile watermarking and blocks with non-regeneratable information are protected using fragile watermarking. Experimental results reveal the accurate identification of type of the block. The performance results reveal that multiple watermarking schemes have reduced the capacity of embedding and consequently improved perceptual quality of the watermarked image.

**Keywords** Multiple watermarking · Intelligent watermarking · Contourlet transforms · Curvelet transforms · Tamper detection · Tamper recovery · Document image

## 1 Introduction

Most of the document images are used as proof of authentication and copyright protection in the business transactions. "Digital watermarking technique has been used as a primary means for copyright protection and integrity management of

K.R. Chetan (✉) · S. Nirmala
Computer Science Department, Jawaharlal Nehru National College of Engineering,
Shivamogga 577201, Karnataka, India
e-mail: chetankr@jnnce.ac.in

S. Nirmala
e-mail: nir_shiv_2002@yahoo.co.in

© Springer Nature Singapore Pte Ltd. 2018
R.K. Choudhary et al. (eds.), *Advanced Computing
and Communication Technologies*, Advances in Intelligent Systems
and Computing 562, https://doi.org/10.1007/978-981-10-4603-2_18

document images [1–3]." The document image consists of information content which can be divided into regeneratable or non-regeneratable blocks. The regeneratable blocks contain minimal changing information content. The blocks having dynamically changing information content are categorized as "non-regeneratable" blocks. Further, there are many empty regions classified as "non-content blocks". Each type of block requires different types of protection. Therefore, there is a need to use multiple watermarking techniques on the different areas of the same document image.

The rest of the paper is organized as follows. Section 2 provides literature review of the existing works. The proposed model is explored in Sect. 3. Experimental results of the proposed scheme are presented in Sect. 4. The performance analysis of the novel technique work is made in Sect. 5. Conclusions of the proposed work are summarized in Sect. 6.

## 2 Literature Review

"Digital watermarking is classified as robust, fragile and semi-fragile based on the robustness to incidental and intentional attacks [4]". A detailed survey of the works on robust, fragile, and semi-fragile watermarking techniques can be found in [5–10]. Houmansadr et al. [11] proposed a watermarking technique based on the entropy masking feature of the Human Visual System (HVS). Kankanhalli et al. [12] developed a watermarking technique by embedding just noticeable watermarks. Radharani et al. [13] designed a content-based watermarking scheme in which watermark is generated using Independent Component Analysis (ICA) for each block of the input image. In [14–16], few works on the segmentation of the image into objects using image statistics and subsequently applying the robust watermarking schemes for each of the objects are described. Shieh et al. [17] used genetics [18] to compute the optimal frequency bands for watermark embedding. Lu et al. [19] developed an algorithm for embedding multiple watermarks into the Vector Quantization (VQ) domain. Sheppard et al. [20] discussed different ways of multiple watermarking like re-watermarking, segmented watermarking, and composite watermarking [20] using different attack scenarios [21, 22].

The literature reviews on the content-based multiple watermarking techniques reveal that most of the existing works lack intelligent application of appropriate watermarking scheme. The previous works on multiple watermarking schemes incur significant degradation in the perceptual quality of the watermarked image. The existing schemes also incur tradeoff between robustness and fragility of the watermarking multiple times. In this paper, a novel intelligent multiple watermarking model is proposed that automatically computes desired type of watermarking for each block of the document image.

# 3 Proposed Model

In this work, a new multiple watermarking model is proposed. The novelty of this approach lies in identifying automatically the type of watermark required for different regions of source document image based on the information content present in that region. This in turn reduces the amount of watermarking to be done in comparison to a single watermarking technique for the entire document image. The proposed model consists of multiple watermarking embedding and extraction process. The input document image is decomposed into blocks of uniform size. To each block, either fragile or semi-fragile watermarking is applied, which is determined automatically. Semi-fragile watermarking is implemented using curvelet-based embedding [23] and fragile watermarking is accomplished using contourlet-based embedding [24]. Extraction process is carried on the blocks of the watermarked image.

## 3.1 Embedding of Multiple Watermarks

The embedding process of multiple watermarks is shown in Fig. 1. Experiments have been conducted exhaustively on all the document images in the corpus to measure the impact of size of the block against accuracy in identifying type of the block. The average number of blocks expected for each type of the block, the number of blocks identified correctly, and processing time are recorded in Table 1. "It can be observed from values in Table 1 that the blocks of lesser size exhibits higher accuracy and consume more time than the blocks of higher dimensions." Considering these parameters size of the block is set to $128 \times 128$. For each block, gradient binarized version of the information content in the block is computed. The type of each block is classified based on the uniformity in distribution and amount of information content present in the block.

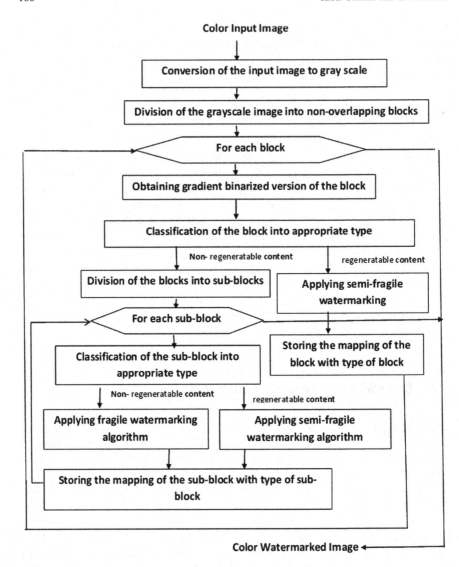

**Fig. 1** Multiple watermark embedding process

**Table 1** Average accuracy in identification of the block type and processing time for identification

| Block size | Non-content | | Uniform content | | Nonuniform content | | Avg IA (in %) | Processing time (in s) |
|---|---|---|---|---|---|---|---|---|
| | EB | IB | EB | IB | EB | IB | | |
| 32 × 32 | 466 | 466 | 210 | 204 | 153 | 150 | 98.97 | 89.12 |
| 64 × 64 | 116 | 116 | 57 | 52 | 33 | 29 | 96.94 | 41.7 |
| 128 × 128 | 25 | 25 | 17 | 15 | 9 | 8 | 95.75 | 23.11 |
| 256 × 256 | 7 | 7 | 4 | 3 | 2 | 2 | 89.61 | 18.16 |

where, *EB* Expected number of blocks evaluated manually by an expert, *IB* Identified number of blocks from the proposed approach, and *IA* Identification accuracy of a block

The gradient binarized version of the information content in the block is computed using the following algorithm:

```
Algorithm to compute gradient binarized version an image:
Input: Image I of size M X N containing gray scale values
(0-255)
Output: Binarized Image of I
Step 1: Decompose I into blocks of size 128 X 128
Step 2: For every pixel p at location (i,j) in the block,
find its gradient value from immediate neighbours i.e.
```

$$g_p(i,j) = \sum_{x=-1}^{1} \sum_{y=-1}^{1} p(x+i, y+j) \tag{1}$$

```
Step 3: Compute maximum gradient value in the block
```

$$g_{max} = \max(g_p) \tag{2}$$

```
Step 4: Set gradient threshold gthresh to (gmax)/2
Step 5: Convert all the pixel values from gray levels to
binary using gthresh. The values of the pixels having gray
levels above gthresh is set to 1 and values of the rest of
the pixels is set to 0.
Step 6: Let n1- number of pixels with pixel value 1 and
n0 - with value 0.
Step 5: Compute new gradient threshold as gthresh1 =
round(n1 * w1 + n0 * w2)
Step 6: If gthresh1 = gthresh , Stop
Otherwise, Set gthresh = gthresh1, Go to Step 4
```

IA is computed using the equation below:

$$IA = \frac{\sum_{block\_type} IB}{\sum_{block\_type} EB}. \tag{3}$$

Experiments are conducted on exhaustive doc-corpus to find the appropriate values of weights $w_1$ and $w_2$. From experimental calculations, it was found setting $w_1 = 0.4$ and $w_2 = 0.6$ leads to a higher degree of accuracy in identification of the type of the block with less amount of time. Hence, the values of weights $w_1$ and $w_2$ are set to 0.4 and 0.6, respectively. The gradient binarized block is classified into an appropriate type using the following algorithm:

```
Algorithm to classify gradient binarized block into vari-
ous block types
Input: Gradient binarized block
Output: Type of the block
Step 1: Computation of the information content of the
block by computing percentage of pixels in the block with
intensity value 1
```

$$IC_b = \frac{\sum_{i=1}^{128X128} g_b}{128X128} \tag{4}$$

```
Step 2: Classifying blocks into content and non-content
blocks
```

$$block\_type = \begin{cases} content, & IC_b > 0.1 \\ non-content, & otherwise \end{cases} \tag{5}$$

```
Step 3: For the blocks of type "content", the cluster
density of rows and columns in the block is computed.
To find row clusters:
    For i = 1 : rows
        For j = 1: cols
            if (g_b(i,j) = g_b(i,j + 1) then
                    continue
            Else
                    row_clusters = row_clusters + 1
            End-if
        End-for
    End-for
Interchanging indices i and j in the above loops will
yield column_clusters
Step 4: avg_clusters=(row_clusters +column_clusters)/2
```

Step 5: $group\_density = \dfrac{avg\_clusters}{128\,X\,128}$

```
Step 6: If group_density exceeds a threshold 0.4, blocks
are sub-divided of same dimension. Steps 3-5 should be
repeated for each subblock. Otherwise, type of the block
is determined using Equation (6).
Step 7: For each content subblock, further classification
into regenerable or non-regenerable content is performed
using following equation:
```

$$block\_content\_type = \begin{cases} regeneratable, & group\_density \le 0.4 \\ non-regeneratable, & otherwise \end{cases} \tag{6}$$

The setting of thresholds (0.1 for information content and 0.4 for group density) is based on the experimental evaluation of the identification accuracy. These thresholds exhibit an accuracy of more than 95% in identification of the type of block. Non-regeneratable blocks are protected using fragile watermarking technique. In this paper an effective fragile watermarking technique based on contourlets [24] is used. Regeneratable blocks are protected using semi-fragile watermarking technique. In this work, semi-fragile watermarking is implemented using Discrete Curvelets Transform (DCLT) [23].

## 3.2 Extraction of Multiple Watermarks

Multiple watermark extraction has similar steps as in multiple watermark embedding process discussed in Sect. 3.1 until the identification of the type of the gradient binarized block. Subsequently, the type of the block/subblock extracted and generated is compared and if there is a mismatch, the corresponding block/subblock of the document image is declared "inauthentic". However, if there is a match, then watermark extraction is carried out based on the type of the block. If the block contains non-regeneratable content, then fragile watermark extraction is performed using contourlets [24]. If the block contains regeneratable content, semi-fragile watermark extraction is performed using DCLT coefficients [23].

## 4   Results

Document images are scanned and a sophisticated corpus is built with different classes like Cheques, Bills, Identity Cards, Marks cards, and Certificates, and each class consists of 30 images. We have tested the accuracy of the identification for all the classes of document images in the corpus. The accuracy values in Table 1 suggest that average accuracy of identification of type of blocks for all classes of document images is more than 95%. Hence, proposed multiple watermarking system exhibits highly accurate identification of type of the block and thus supports for application of intelligent multiple watermarking. Figure 2 depicts that watermarked image is perceptually similar to source document image in the corpus. An example of insertion attack on a regeneratable block and modification attack on a non-regenerable block of the watermarked image is illustrated in Fig. 2. Semi-fragile watermarking extraction results reveal that there is a great degree of accuracy in tamper detection. Further, accurate tamper recovery of the nonuniform content block is also observable in Fig. 2.

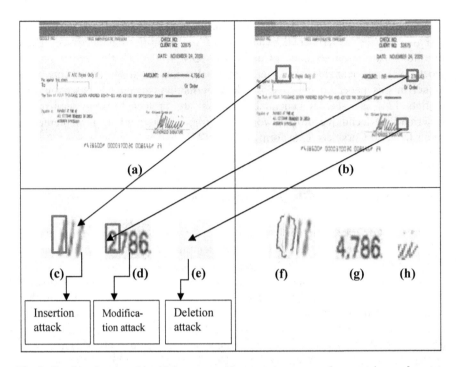

**Fig. 2** Results of proposed multiple watermarking system **a** source document image, **b** watermarked image, **c** zoomed up uniform content block tampered with insertion attack, **d** zoomed up nonuniform content block containing preprinted information content with modification attack, **e** zoomed up nonuniform content block containing handwritten information content with deletion attack, **f** tamper detection results of uniform content block, **g** and **h** tamper recovery results of nonuniform content blocks

## 5 Analysis

The proposed watermarking system is measured for performance in terms of the following parameters: (i) Fidelity analysis using Peak Signal-to-Noise Ratio (PSNR), (ii) Accuracy of Tamper detection, and (iii) Accuracy of Tamper recovery.

### 5.1 Fidelity Analysis

The fidelity of the proposed multiple watermarking scheme is evaluated in terms of PSNR [25]. A plot of PSNR values is shown in Fig. 3 for different classes of the document images. The graph shown in Fig. 3 reveals that PSNR values of the multiple watermarking schemes are better than semi-fragile and fragile watermarking schemes when applied separately. The amount of watermarking performed

**Fig. 3** PSNR values of
different classes of document
images in the corpus

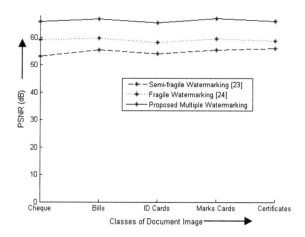

**Table 2** Average TDA and TRA values for different intentional attacks

| Intentional attacks | Existing semi-fragile watermarking scheme [23] | Existing fragile watermarking scheme [24] | Proposed multiple watermarking scheme | |
|---|---|---|---|---|
| | TDA | TRA | TDA | TRA |
| Insertion | 0.9 | 0.87 | 0.94 | 0.92 |
| Deletion | 0.92 | 0.91 | 0.95 | 0.94 |
| Modification | 0.87 | 0.87 | 0.94 | 0.92 |

depends on the type of the block. Hence, the noise induced due to watermarking is reduced and this contributes for the better fidelity of the watermarked image.

## 5.2 Effectiveness Analysis of Detection and Correction Operations

The effectiveness in detecting and correcting tamper of an attacked block of the proposed multiple watermarking schemes is evaluated in terms of accuracy of tamper detection and tamper recovery parameters. Accuracy of tamper detection and recovery is evaluated as follows:

$$\text{TDA} = 1 - \frac{\sum_{i=1}^{n}(ta_i \oplus td_i)}{n}, \quad \text{TRA} = 1 - \frac{\sum_{i=1}^{n}(ta_i \oplus tr_i)}{n}, \quad (7)$$

where $n$—total number of bits in the fragile watermarked blocks, $ta$—tampered bit, and $td$—tamper detection bit. The average values of TDA and TRA are computed for all document images in the corpus under different intentional attacks for proposed multiple watermarking scheme and existing semi-fragile [23] and fragile watermarking [24] schemes separately. These values are tabulated in Table 2. It can be observed that proposed multiple watermarking scheme exhibits better performance in both detection and correction operations on the tampered information content of document image.

# 6 Conclusions

A novel watermarking technique for protection of document images using multiple watermarking schemes on the same image is proposed in this paper. The blocks of a document image have been automatically classified into various types with greater accuracy. The performance analysis of the proposed approach reveals significant improvement in the fidelity of the watermarked image. The proposed scheme also outperforms the existing methods [23, 24] with better tamper detection and recovery capabilities. Improvement on the accuracy of identification of the type of block is the task of further enhancement.

# References

1. Wu, M., Liu, B.: Watermarking for image authentication. In: Proceedings of the IEEE International Conference on Image Processing, pp. 437–441 (1998)
2. Cox, I., Miller, M., Bloom, J., Fridrich, J., Kalker, T.: Digital Watermarking and Steganography. Morgan Kaufmann Publishers Inc., San Francisco (2007)
3. Hartung, F., Kutter, M.: Multimedia watermarking techniques. Proc. IEEE 87(7), 1079–1107 (2002)
4. Potdar, V.M., Han, S., Chang, E.A.: A survey of digital image watermarking techniques. In: 3rd IEEE International Conference on Industrial Informatics, pp. 709–716 (2005). doi:10.1109/Indin.2005.1560462
5. Mirza, H., Thai, H., Nakao, Z.: Color image watermarking and self-recovery based on independent component analysis. Lect. Notes Comput. Sci. 5097, 839–849 (2008)
6. Wang, M.S., Chen, W.C.: A majority-voting based watermarking scheme for color image tamper detection and recovery. Comput Stand. Interfaces 29, 561–571 (2007)
7. Bas, P., Chassery, J.M., Macq, B.: Geometrically invariant watermarking using feature points. IEEE Trans. Image Process. 11(9), 1014–1028 (2002)
8. Qi, W., Li, X., Yang, B., Cheng, D.: Document watermarking scheme for information tracking. J. Commun. 29(10), 183–190 (2008)
9. Dawei, Z., Guanrong, C., Wenbo, L.: A chaos-based robust wavelet-domain watermarking algorithm. Chaos, Solitons Fractals 22(1), 47–54 (2004)
10. Schirripa, G., Simonetti, C., Cozzella, L.: Fragile digital watermarking by synthetic holograms. In: Proceedings of European Symposium on Optics/Photonics in Security & Defence, pp. 173–182 London (2004)
11. Houmansadr, A., et al.: Robust content-based video watermarking exploiting motion entropy masking effect. In: Proceedings of the International Conference on Signal Processing and Multimedia Applications, pp. 252–259 (2006)
12. Kankanhalli, M.S., Ramakrishnan, K.R.: Adaptive visible watermarking of images. In: IEEE International Conference on Multimedia Computing and Systems, vol. 1, pp 568–573 (1999)

13. Radharani, S., et al.: A study on watermarking schemes for image authentication. Int. J. Comput. Appl. (0975–8887), **2**(4), 24–32 (2010)
14. Kay, S., Izquierdo, E.: Robust Content Based Image Watermarking. In: Proceedings of Workshop on Image Analysis for Multimedia Interactive Services (2001)
15. Kim, M., Lee, W.: A content-based fragile watermarking scheme for image authentication. Lect. Notes Comput. Sci. Content Comput. **0302**, 258–265 (2004)
16. Habib, M., Sarhan, S., Rajab, L.: A robust fragile dual watermarking system in the dct domain. Lect. Notes Comput. Sci. Knowl.-Based Intell. Inf. Eng. Syst. **3682**, 548–553 (2005)
17. Shieh, C.S., et al.: Genetic watermarking based on transform-domain techniques. J. Pattern Recogn. **37**, 555–565 (2004)
18. Goldberg, D.E.: Genetic Algorithms in Search Optimization and Machine Learning. Addison-Wesley, Reading, MA (1992)
19. Lu, Z.M., Xu, D.G., Sun, S.H.: Multipurpose image watermarking algorithm based on multistage vector quantization. IEEE Trans. Image Process. **14**(6), 822–831 (2005). doi:10. 1109/Tip.2005.847324
20. Sheppard, N.P., Safavi-Naini, R., Ogunbona, P.: On multiple watermarking. In: Dittmann, J., Nahrstedt, K., Wohlmacher, D. (eds.) Multimedia and Security: New Challenges Workshop, p. 38871 (2001)
21. Voloshynovskiy, S., Pereira, S., Pun, T., Eggers, J.J., Su, J.K.: Attacks on digital watermarks: classification, estimation based attacks, and benchmarks. IEEE Commun. Mag. **39**(8), 118–126 (2001)
22. Zhang, X., Wang, S.: Watermarking scheme capable of resisting sensitivity attack. IEEE Sign. Process. Lett. **14**(2), 125–128 (2007)
23. Chetan, K.R., Nirmala, S.: An efficient and secure robust watermarking scheme for document images using integer wavelets and block coding of binary watermarks. J. Inf. Sec. Appl. **24–25**, 13–24 (2015)
24. Chetan, K.R., Nirmala, S.: A novel fragile watermarking scheme based on contourlets for effective tamper detection, localization and recovery of handwritten document images, IEEE Sign. Process. Lett. (Communicated)
25. Aggarwal, D.: An efficient watermarking algorithm to improve payload and robustness without affecting image perceptual quality, J. Comput. **2**(4) (2010). ISSN 2151-9617

# Part V
# Advanced Computing: Big Data

# Research of Access Optimization of Small Files on Basis of B + Tree on Hadoop

Yan Wang, YuQiang Li, YuWen Li, Yilong Shi and Weiwei Li

**Abstract** Hadoop, the open-source software for reliable, scalable, distributed computing used in the processing and storage of extremely large data sets, is originally designed to store large amounts of large files resulting in huge wastage of storage space for Data Node and increase in the memory space utilization, for Name Node, when dealing with massive small files. For the above shortcomings, this paper puts forward a optimization design for small files access scheme, which speeds up the small file location through the file index, on the Hadoop platform based on B + tree index, resulting in the improvement in the access efficiency of small files. The effectiveness of the proposed scheme is experimentally validated.

**Keywords** Hadoop · HDFS · B + tree · Small files · Sequence file

## 1 Introduction

In recent years, with the continuous advances in information science and technology, "big data" technology has gradually become the focus of attention for both industry and academia [1, 2].

Faced with Internet data presented this explosive growth [3], Google company publishes papers which first proposed the GFS, MapReduce, and other distributed data processing technology to deal with these massive data in 2003 [4, 5]. Thereafter, Apache Foundation develops a distributed system Hadoop.

Since Hadoop development platform is originally intended to store large amounts of large files, which causes huge waste of storage space for DataNode, increased memory space utilization for NameNode when dealing with massive small files. Therefore, improving the processing ability of small HDFS file is paid more and more attention by the outside world. At present, there is a lot of research

Y. Wang (✉) · Y. Li · Y. Li · Y. Shi · W. Li
School of Computer Science and Technology, Wuhan University of Technology, Wuhan, China
e-mail: wy_hult@163.com

© Springer Nature Singapore Pte Ltd. 2018
R.K. Choudhary et al. (eds.), *Advanced Computing and Communication Technologies*, Advances in Intelligent Systems and Computing 562, https://doi.org/10.1007/978-981-10-4603-2_19

197

work on Hadoop small file access method both in academic research and in the application of the major Internet Division.

Mackey et al. [6] put forward a scheme that assigning a quota for each client in the HDFS file system and verifying the effectiveness of it. Then, on the basis of the solutions of Grant Mackey et al., Vorapongkitipun and Nupairoj [7] put forward a new Hadoop Har (archive) scheme to solve the problem that the metadata information of small files in HDFS system occupy the memory of NameNode, and improve the metadata memory utilization of the NameNode and the access efficiency of small files from two aspects.

In China, many researchers and Internet companies, like taobao, tencent, etc., put forward their own solutions for the small file access problems. Changtong [8] merges small documents into one big file, and builds a Hash index for each of the merged files to improve the efficiency of small file access. Focusing on the problem that huge numbers of small files impose heavy burden on NameNode of HDFS, Chen et al. [9] present a kind of Cache strategy to improve the reading efficiency of small files on HDFS. In 2012, Zhang and Liu [10] puts forward a new scheme; their core idea is to use the IO multiplexing reactor model to deal with a large number of request tasks of small files and constantly creates indexes for the merged files. But for different types of files, they do not make effective consideration. Then, on the basis of the research of Yang et al. [11] team puts forward the mechanism that takes a different file merging and prefetching for different sizes of the file to deal with small file which increases the read and write performance by 70%.

It notes the impact of the access of small files that it brings to Hadoop platform, so Hadoop itself proposed three small file merger proposals, Hadoop Archive, SequenceFile, and CombineFileInputFormat. But the SequenceFile consolidation program uses a sequential manner to search small files after merger, which is inefficiencies and cannot locate the file quickly under the condition of large amount of files, and seriously affects read performance of small files.

This paper establishes a B + tree index for the merged file on the basis of SequenceFile, to speed up the small file location through file index, so as to improve the reading efficiency of small files.

## 2   The Search for SequenceFile Based on B + Tree

### 2.1   The Construction of B + Tree Index

Combining SequenceFile merging scheme with the construction of B + tree index strategy, a concrete design idea is given: due to the massive small image files according to the years before the classified into multiple SequenceFile based large files, so for the design of the B + tree index, would require the two levels of index of the search process. First of all, build the B + tree index for all small files with the corresponding SequenceFile large files, then the small files can be quickly located to its large files by small file name. Then, build B + tree index for each large

SequenceFile, so specific information can be located by the large file rapidly and restored the files. Here, in this article, the B + tree index for all the small files and the corresponding SequenceFile large files is called primary index, and the B + tree index for each large SequenceFile is called secondary index.

### 2.1.1 The Design of the Primary Index Structure

The primary B + tree index structure is shown in Fig. 1. To combine with massive small images of the data, it is assumed that there are six pictures called a.jpg, b.jpg, c.jpg, d.jpg, e.jpg, and f.jpg, and they are classified to the large file of the corresponding year according to the year. So the pictures called a.jpg, c.jpg, and f.jpg are classified to the file called 2014, and the pictures called b.jpg, d.jpg, and e.jpg are classified to the file called 2015. Here, the order of B + tree is three, namely when the key word in the node is more than three, the node splits.

For primary B + tree index structure, the B + tree index structure design is as follows:

```
Public class  BplusTree{
    protected Node root;        /*root node*/
    protected int m;            /*order m */
    protected Node head;        /*head of the leaf nodes*/
}
```

The structural of a single node in the B + tree is designed as follows; among them, the key word entry node is designed as set type to store key information in each node. Each keyword information is set as the generic type Entry $\langle K, V \rangle$. Entry $\langle K, V \rangle$ represents an entity of the Map that is a $\langle Key, Value \rangle$ pair. In the implementation of primary B + tree index structure, the K corresponds to each picture filename which is a String type, and V corresponds to the classified SequenceFile name which is Object type.

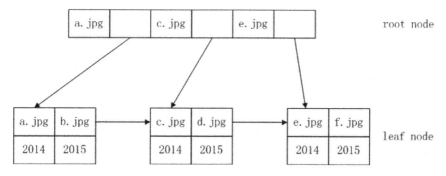

**Fig. 1** The figure of primary index structure

```
Public class Node{
  Protected Boolean isLeafNode;
  Protected Boolean isRootNode;
  Protected Node parentNode;
  Protected Node previousNode;
  Protected Node nextNode;
  Protected List<Entry<String,Object>> entries;
  Protected List<Node> children;
}
```

### 2.1.2  The Structural Design of the Secondary Index

The small files can be quickly located to its large files by small file name through the primary index. However, the secondary index that the B + tree index for each large SequenceFile is needed to find the specific small file information in SequenceFiles. The secondary index structure is shown in Fig. 2. When the picture c.jpg is searched, it is found that c.jpg is in the file named 2014 and there are four other pictures a.jpg, f.jpg, h.jpg, and m.jpg.

For secondary B + tree index structure, the structure is same as the primary B + tree index. The realization is seen in Sect. 2.1.1. However, K and V represent different information; the K corresponds to the image name in the merged SequenceFile files, and V corresponds to the binary of each image in merged SequenceFile files. Because each image binary is larger, it is inconveniently depicted in the figure, and therefore, the binary code of images is replaced with the value.

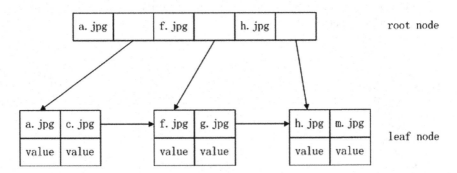

**Fig. 2**  The figure of the secondary index structure

## 2.2 The Optimization of Small File Read Scheme Based on B + Tree

If the B + tree index file is stored in the NameNode node, it is undoubtedly exacerbated the NameNode node's workload. At the same time, if the NameNode node fails, the Hadoop cluster will also stop work, so the B + tree index file stored in the NameNode node will be lost. So, in conclusion, this paper puts forward a optimization design idea of small file reading scheme on the Hadoop platform based on B + tree index: Secondary B + tree index file is stored in the DataNode node in order to reduce disk I/O access to the DataNode node number and speed up the small file search. The improved HDFS file reading process chart is shown in Fig. 3.

The improved part is shown in the dotted box in Fig. 3. The improved accessing process is as follows:

Step 1: The client initiated the RPC () request to the remote NameNode.
Step 2: After receives the request from the client, the NameNode returns the DataNodes, the index file Block corresponds to through the mapping relationship between B + tree index file and the Block and the mapping relationship between the Block and DataNodes, stored by the NameNode.
Step 3: The client selects the nearest DataNode to read Block in B + tree index file.
Step 4: If find the needed file through the index file Block in current DataNode, then read the binary of the file and reduction it. If do not find the needed file in current DataNode, cancel connections with current DataNode node and read the next index file Block to find the most suitable DataNode.
Step 5: After finishing reading the whole index files, the client closes the whole file system.

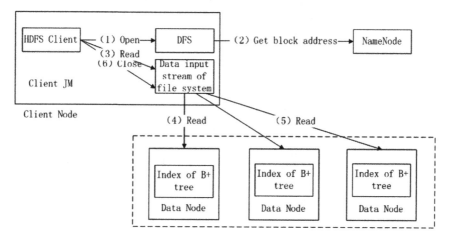

**Fig. 3** The figure of improved HDFS file reading process

# 3 The Experiment

## 3.1 Experimental Environment

The experiment need to set up a Hadoop cluster; in this paper, the Hadoop cluster consists of four lenovo M4350 business machines: One of them as the NameNode, and the other three as a DataNode. Hardware configuration and software configuration of the computers in cluster are shown in Table 1.

Detail name and IP address configuration are shown in Table 2.

## 3.2 Experimental Setup

The contrast experiment is carried out on the Hadoop—version 1.0.3 system. The size of the data block is set to 64 MB and parameter of copies is set to 3 in the experiment. The size of picture type small files is between 100 KB and 15 MB. The size of document type small files is between 100 KB and 15 MB.

This paper will compare the read time and memory utilization of the NameNode for two kinds of small file, and test two kinds of search strategy when a number of small files in the are 50, 100, 500, 1000, 5000, and 10,000.

## 3.3 Results and Analysis

### 3.3.1 Read Time of Small File

The contrast curve of reading time is given as follows (Fig. 4).

According to the shape of the curve in the figure, in view of the image type small files, the smaller the number of small files is, the better the MapFile index reading method is; the bigger the number of small files is, the better the B + tree index

Table 1 The table of computer configuration

| Hardware configuration | CPU | Memory | Hard disk |
|---|---|---|---|
| | I3 3240 | 4G | 320G |
| Software configuration | OS | JDK | Hadoop |
| | Ubuntu12.04 (64bit) | JDK1.6.0_45 | Hadoop1.0.3 |

Table 2 The table of IP addresses

| The name of the machine | Master | Slave45 | Slave46 | Slave47 |
|---|---|---|---|---|
| IP address | 192.168.1.144 | 192.169.1.145 | 192.168.1.146 | 192.169.1.47 |

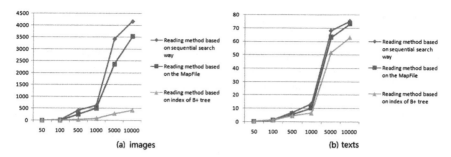

**Fig. 4** The figure of the reading time

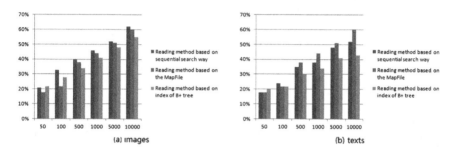

**Fig. 5** The figure of the memory utilization

reading method is better. At the same time, according to the comparison results in the table, the reading time using the search way based on the B + tree index after merger based on the SequenceFile merger plan is 80% (text type was 30%) less than the original sequential search way and the way based on the index of MapFile. The experimental results also verify again that for images (small files) type of massive small files on the reading efficiency; this proposed small file reading scheme based on B + tree index is efficient.

### 3.3.2   The Memory Utilization of NameNode

The contrast histogram of NameNode memory utilization is given as follows (Fig. 5).

According to the histogram, in view of the memory utilization of the NameNode node, the smaller the number of small files is, the better the MapFile index reading method is; the bigger the number of small files is, the better the B + tree index reading method is. The experimental results also verify again that for images (small files) type of massive small files on the reading efficiency, this proposed small file reading scheme based on B + tree index is efficient.

# 4 Conclusion

This paper first analyzes the shortages of SequenceFile merging scheme on Hadoop platform, and achieves the small files search by integrating small files into big files with certain rules according to the characteristics of the experimental data. SequenceFile merging scheme uses a sequential search to read small files which seriously influences the small file reading efficiency. Therefore, this paper presents a design scheme based on B + tree index file, and designs and achieves the structure and function of B + tree index. Then, analyze the original HDFS file read process deeply and find out the problems which need to be modified. Finally, the improved scheme is applied to the Hadoop platform. Simulation experiment establishes that the proposed reading scheme based on B + tree index file is feasible and effective in improving the efficiency of small file access.

# References

1. Xiao, F., Li-Lei, Q.I.: Exploration of big data processing technology. J. Comput. Mod. **1**, 75–77 (2013)
2. Meng, X., Xiang, C.: Big data management: concepts, techniques and challenges. J. Comput. Res. Dev. **1**, 146–169 (2013)
3. Wang, Y.Z.: Network big data: present and future. J. Chin. J. Comput. **6**, 1125–1138 (2014)
4. Jiang, L., Li, B., Song, M.C.: The optimization of HDFS based on small files. In: 3rd IEEE International Conference on Broadband Network and Multimedia Technology, pp. 912–915. IEEE Computer Society, Beijing (2010)
5. Zhang, Y., Liu, D.C.: Improving the efficiency of storing for small files in HDFS. In: 2012 International Conference on Computer Science and Service System, pp. 2239–2242. IEEE Computer Society, Jiangsu (2012)
6. Mackey, G., Sehrish, S., Wang, J.C.: Improving metadata management for small files in HDFS. In: 2009 IEEE International Conference on Cluster Computing and Workshops, pp. 1–4. Institute of Electrical and Electronics Engineers Inc., New York (2009)
7. Vorapongkitipun, C., Nupairoj, N.C.: Improving performance of small-file accessing in Hadoop. In: 11th International Joint Conference on Computer Science and Software Engineering, pp. 200–205. IEEE Computer Society (2014)
8. Changtong, L.: An improved HDFS for small file. In: 18th International Conference on Advanced Communications Technology, pp. 474–477. Institute of Electrical and Electronics Engineers Inc., New York (2016)
9. Chen, J., Wang, D., Fu, L., Zhao, W.: An improved small file processing method for HDFS. Int. J. Digit. Content Technol. Appl. **6**, 293–304 (2012)
10. Zhang, Y., Liu, D.C.: Improving the efficiency of storing for small files in HDFS. In: 2012 International Conference on Computer Science and Service System, pp. 2239–2242. IEEE Computer Society, Washington (2012)
11. Zhang, S., Miao, L., Zhang, D., et al.: A Strategy to deal with mass small files in HDFS. In: 6th International Conference on Intelligent Human-Machine Systems and Cybernetics, pp. 331–334. Institute of Electrical and Electronics Engineers Inc., New York (2014)

# Part VI
# Advanced Computing: Machine Translation

# English to Bodo Phrase-Based Statistical Machine Translation

Md. Saiful Islam and Bipul Syam Purkayastha

**Abstract** In spite of its inclusion in the scheduled languages of India and being one of the official languages of Assam, significant research and work is yet to be reported for machine translation of Bodo. The primary objective of the proposed system in the paper, is to develop an English to Bodo Phrase-Based Statistical Machine Translation (SMT) system using Moses, and Tourism domain English to Bodo parallel corpora. The performance of the proposed system using the BLEU score is 65.09.

**Keywords** Bodo language · English language · Machine translation · Moses

## 1 Introduction

In this section, we discuss briefly about the Machine Translation and natural languages which are used in the proposed system.

### 1.1 Machine Translation

Machine Translation is defined as a system that automatically translates text from a source natural language (SL) to target natural language (TL) using computers. It produces translation between two particular languages and it may be either unidirectional or bidirectional [1]. The idea of machine translation was originated by the philosopher René Descartes in the seventeenth century [2]. 'Machine Translation using digital computers has been a grand challenge for computer scientists,

Md. Saiful Islam (✉) · B.S. Purkayastha
Department of Computer Science, Assam University, Silchar, Assam, India
e-mail: sislam.mca@gmail.com

B.S. Purkayastha
e-mail: bipul_sh@hotmail.com

© Springer Nature Singapore Pte Ltd. 2018
R.K. Choudhary et al. (eds.), *Advanced Computing
and Communication Technologies*, Advances in Intelligent Systems
and Computing 562, https://doi.org/10.1007/978-981-10-4603-2_20

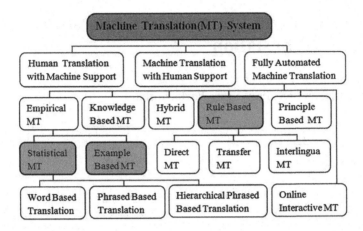

**Fig. 1** Approaches of MT

mathematicians and linguists since the first international conference was held at MIT in 1952'. Nowadays, it is a challenging research problem, universally, in the areas of Computational Linguistics and Natural Language Processing (NLP). Machine Translation systems have different approaches. The different approaches of MT systems are shown in Fig. 1.

## 1.2 Bodo Language

Bodo (pronounced as Boro), spoken by approximately 1.4 million people of North-East India and Nepal [3] is one of the famous languages of North-East India. Bodo, also known as Mech, is a language of the Tibeto-Burman Branch of Sino-Tibetan Language. It is the co-official language of Assam (BTC-Bodoland Territorial Council) and is a recognized language in India. The Bodo language is mainly used by the population of Kokrajhar, Chirang, Baksa and Udalguri districts of Assam. It is also spoken by some people of Cooch Behar, Alipurduar and Jalpaiguri districts of West Bengal, as well as in Meghalaya. The Bodo language is written in Latin and Devanagari scripts.

## 1.3 English Language

English was the first spoken language in early mediaeval England and is now a global lingua franca [4]. The English language is spoken mainly by the population of Australia, Canada, Ireland, New Zealand, United Kingdom and the United States. It is an official language of almost sixty sovereign states and third most common native language in the World. English was introduced in India (1830) during the East India Company regime. The Constitution of India (1951) declared Hindi as the primary official language and English as the associate official language of India. Now, it is the third most spoken language in India. It has various dialects in India due to the influence of local languages.

## 2 Related Works

In this section, developments of the MT system using Statistical Machine Translation approach is briefly reviewed.

### 2.1 SMT System in the World

A large number of MT works have been done by several institutions/organizations in several countries on natural languages using SMT approach. In 1949, Warren Weaver introduced the first concepts of SMT approach [5]. In this paper, we describe some of the most recent MT works where SMT approach is used. The main works of SMT system were carried out by the researchers at IBM. First, they used the SMT approach in the Candide project at IBM in 1988 [6]. The Candide project was tested using large amount of parallel text corpora of French and English languages. Now, the SMT approach has become the most popular one of the MT approaches for many MT research groups. The EuroMatrix project began to develop MT systems using SMT approach between all the European Union languages in 2006 [6]. Recently, the Phrase-Based SMT is highly successful technique and used by many MT research groups. French to English Statistical Phrase-Based Machine Translation was developed by Philipp Koehn using Moses, University of Edinburgh [7, 8]. English to Bangla Phrase-Based SMT was developed by Md. Zahurul Islam, Saarland University, Germany, 2009 [9]. English to Urdu Hierarchical Phrase-Based SMT was developed by Nadeem Khan et al., Pakistan. The Google Translate is developed by Google using the SMT approach to translate texts of various languages in 2006. The Bing Translate is developed by Microsoft using the SMT approach to translate texts of various languages in 2009 [10].

### 2.2 SMT System in India

Lots of MT work has been reported using the SMT approach in India. The institutions like CDAC, TDIL, MCIT(Ministry of Communications and Information Technology), IIT, NIT, and the Universites have developed some Machine Translation systems for Indian natural languages using SMT approach. The projects ANUVAADAK at IIT Bombay, E-ILMT (Consortium of Nine Institutions, 2006), and Shakti [11], a few notable examples were developed using the SMT approach in India [2, 12]. Some other examples of SMT system are mentioned below:

- Assamese to English SMT system integrated with transliteration was developed by P. Das and K.K. Baruah, Gauhati University, 2014 [13].
- English to Assamese MT system was developed by Moirangthem Tiken Singh, Sourav Gohain and Rajdeep Borgohain, Dibrugarh University, 2014 [14].
- English to Hindi SMT system was implemented by Nokul Sharma and Parteek Bhartia, Thapar University, 2011 [10].
- English to Malayalam SMT system was developed by A. George, A.S. College of Engineering and Technology, 2013 [15].
- English to Kannada SMT system was developed by P.J. Antony, P. Unnikrishnan and K.P. Soman, 2010 [2].
- English to Telugu Statistical Machine Translation system was implemented by A. Nalluri and V. Kommaluri, 2011 [2].

## 3   Tools for Implementing the Proposed System

In this section, the tools which were used to implement the proposed system are briefly discussed.

### 3.1   Kali Linux

Kali Linux is basically an operating system (OS). It belongs to Debian-Based Linux distribution and is designed for advanced penetration testing and security auditing. The Kali Linux was released in the month of March 2013. The Linux is the primary development platform for Moses.

### 3.2   Moses

Moses, an open source SMT toolkit was developed at Edinburgh University by Philipp Koehn and Hieu Hoang in 2005 [16]. It uses a huge amount of parallel corpora to train the system and is widely used in the Phrase-Based SMT framework. Moses can run on both Linux and Windows OS (under Cygwin).

## 3.3 IRSTLM and KenLM

IRSTIM and KenLM both are Language Model toolkits, used for estimation, representation and computation language models in the SMT system. The Language Model is developed only for the target language using the KenLM and IRSTLM toolkit [16]. IRSTLM is freely available on online and KenLM comes with Moses.

## 3.4 GIZA$^{++}$

GIZA$^{++}$, an SMT toolkit, used to build translation modelling in the Phrase-Based SMT system, designed by Franz Josef Och [16]. It is also used for word alignments in a bilingual parallel corpus to train the translation modelling system. It is freely available online.

## 3.5 Corpus

Corpus is a defined collection of text of some specific languages in digital format. The scope of the corpus is very vast in NLP for linguistic analysis. We have collected tourism domain English to Bodo parallel text corpus from Technology Development for Indian Languages (TDIL) to implement the proposed system. The length of the parallel corpus is 9500 sentences of each language.

# 4 System Implementation

In this section, the approach and procedures are discussed to implement the proposed MT system. We have used Phrase-Based SMT approach to develop the system.

## 4.1 Statistical Machine Translation

The Statistical Machine Translation is one of the most commonly used approaches of MT system where translations are done on the basis of statistical models. It comes under Empirical (or Corpus-Based) MT system and uses enormous amount of bilingual parallel text corpora in the source and target languages to achieve high quality translation result [8]. The SMT approach offers the best solution to ambiguity problem. The main advantages of SMT approach are, it is easy to build and

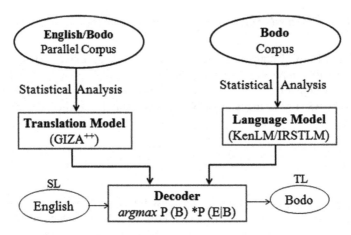

**Fig. 2** Architecture of English to Bodo SMT

maintain, no linguistic knowledge required and reduces the human effort [10]. There are three types of SMT approach which are shown as below:

- Word-Based SMT
- Phrase-Based SMT
- Hierarchal Phrase-Based SMT.

The Statistical Machine Translation approach contains three different models which are shown as below:

- **Language Model (LM)**: The language model computes the probability of the target language (Bodo language) $B$, i.e. $P(B)$.
- **Translation Model (TM)**: The translation model helps to compute the probabilities of the source language sentence $E$ (English) and a target language sentence $B$ (Bodo), i.e. $P(E|B)$.
- **Decoder**: The decoder maximizes the probability of the product of LM and TM, i.e. argmax $P(B) * P(E|B)$.

The architecture of English to Bodo Statistical Machine Translation is mentioned in Fig. 2.

## 4.2 Phrase-Based SMT

A phrase is a sequence of consecutive words or sentence. The Phrase-Based SMT approach is an accurate and widely used approach in the MT system nowadays. There are several advantages of Phrase-Based Translation than Word-Based Translation. It can capture local context and uses that context in the translation. This approach allows the translation of non-compositional phrases and can handle many to many translations. Phrase translations are learned from data in an unsupervised way. In Phrase-Based Translation, each sentence of source and target languages are divided into different phrases before translation. The alignment follows certain patterns between the phrases in the source and target sentences which is almost like Word-Based Translation [7, 8].

In Phrase-Based SMT, the following steps are performed to implement the proposed system using Moses and Perl language.

### 4.2.1 Corpus Preparation

To train the Phrase-Based SMT system we need large amount of parallel corpora which are aligned at the sentence level [16]. In the proposed system, tourism domain English to Bodo parallel text corpus is used to train the translation system. We have used 9500 parallel sentences of each English and Bodo language. To train the system, two separate gedit files are created in UTF-8 format for English and Bodo corpus on Kali Linux. Now, the following steps are done to prepare the English and Bodo corpus for training the translation system.

- **Tokenization**: Tokenization is prepared for both the English and Bodo text corpora. It is used to insert space between words and punctuation in the corpora.
- **True casing**: True casing is used to convert the first words of each sentence to their most probable casing for both the English and Bodo tokenized corpora.
- **Cleaning**: Cleaning is used for removing the long sentences, empty sentences and extra spaces from both the English and Bodo corpora.

### 4.2.2 Language Model

The language model is used to measure the fluency of the output sentence in the proposed system. The LM is built only for the target language (in this system, Bodo language). In the proposed system, the LM toolkit, KenLM is used for training the phrase-based language modelling. The IRSTLM toolkit is also used separately to develop the language modelling. The goal of LM is to calculate the probability of sentences of target language $P(B)$. The LM calculates the probability of the sentences of target language using the n-gram model.

An n-gram is an adjacent series of n elements from a given series of text. The n-gram language model is based on statistics of how probably the words are to follow each other. The LM decomposes the probability of a target sentence as the probability of particular words $P(w)$ using Markov Chain Rule [14] as shown in Eq. (1).

$$\begin{aligned} P(B) &= P(w_1, w_2, w_3, \ldots, w_n) \\ &= P(w_1)P(w_2|w_1)P(w_3|w_1w_2)P(w_4|w_1w_2w_3)\ldots P(w_n|w_1w_2\ldots w_{n-1}) \end{aligned} \tag{1}$$

The n-gram language model is used to make less complex the task of approximating the probability of a word given all the previous words. In n-gram model, the size $N = 1$ is represented as unigram, size $N = 2$ is represented as bigram (or digram), size $N = 3$ is represented as trigram, and size $N = n$ is known as n-gram. The n-gram probabilities can be computed in a straightforward manner from a corpus. The formula for calculating trigram probabilities is given in Eq. (2).

$$P(w_n|w_{n-2}, w_{n-1}) = \frac{\text{count}(w_{n-2}w_{n-1}w_n)}{\text{count}(w_{n-2}w_{n-1})} \qquad (2)$$

Here, count $(w_{n-2}w_{n-1}w_n)$ indicates the number of occurrences of the sequence $w_{n-2}w_{n-1}w_n$ in the corpus.

In the proposed system, 3-gram language model is used to build the language model.

### 4.2.3 Translation Model

The translation model calculates the probability of the source sentence ($E$) of English language for a given target sentence ($B$) of Bodo language, i.e. $P(E|B)$. The TM computes the probability by considering the behaviour of the phrases. To train the TM, the most crucial and indispensable step is word (or phrase) alignment. Giza$^{++}$ is used for word alignment in TM. A word alignment between sentences tells us exactly how each word in sentence $E$ is translated in $B$. Since, the computation of TM probabilities is not possible at sentence level, therefore the process is broken down into small units for words or phrases and their probabilities are learned [13]. The following Fig. 3 is an example of English to Bodo Phrase-Based TM.

### 4.2.4 Decoder

The task of decoder is to maximize the probability of the translated text in SMT system. The Moses decoder performs decoding which is the process of translating a source sentence into target translated sentence using LM and TM. In the proposed system, Moses decoder accepts English language text as input and generates Bodo language text as output. The decoder uses $A^*$ search based on heuristic search method to find the best possible translation [16]. The decoder can maximize the

**Fig. 3** English to Bodo Phrase-Based TM

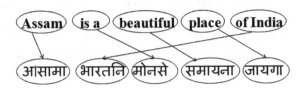

translation probability using the output obtained from the LM and TM. Finally, the decoder gives the final translation result using the following function:

$$P(E, B) = \arg\max P(B) * P(E|B).$$

## 5 Results and Evaluation

In this section, the results and the quality of the translation of the proposed system is discussed.

### 5.1 Result

The English to Bodo Phrase-Based SMT system is tested with different numbers of tourism domain parallel sentences of English and Bodo languages for translation result. It has been observed that if we increase the number of sentences in the given parallel corpora to train the system, then the quality of the translation result is also enhanced. Lastly, tourism domain English to Bodo parallel corpora with 9500 sentences of each language is used to train the proposed system. The following command is used to run the Moses for translation result:

```
~/mosesdecoder/bin/moses -f ~/working/mert-work/moses.ini
<~/corpus/training/input.tourism.eng-bod.en > output.tourism.eng-bod.bd
```

Here, input.tourism.eng-bod.en is an input English text file and output.tourism. eng-bod.bd is an output Bodo text file.

The translation results for ten English and Bodo parallel sentences using the proposed system are shown in Table 1.

### 5.2 Evaluation

Bilingual Evaluation Understudy (BLEU) technique, 'an algorithm for evaluating the quality of text which has been machine-translated from one natural language to another', is applied to evaluate the translation quality. The technique developed by Kishore Papineni et al. in 2001 [16, 17] is a very useful for automatic evaluation of MT systems and is universally accepted. It depends on the average of matching n-grams between a proposed translation and a reference translation. The BLEU

**Table 1** English to Bodo Translation Result

| INPUT [English Sentences] | OUTPUT [Bodo Sentences] |
|---|---|
| Assam is a beautiful place of India | आसामा भारतनि मोनसे समायना जायगा। |
| Jaipur is the capital of Rajasthan | जयपुरा राजस्थाननि राजथावनि। |
| Taj Mahal was built by Shah Jahan | ताज महलनि शाहजाहाना लुर्दोमोन। |
| Agra is best known for the Taj Mahal | आगराया ताज महलनि थाखाय साबसिन मिथिजानाय। |
| I love green tea | आं सोमखोर साहा मोजां मोनो। |
| Assam is famous for one-horn rhinoceros | आसामा थसे गंगोनां गान्दानि थाखाय मुंदांखा। |
| Kaziranga park is very beautiful | काजिरङा हायग्रामाया जोबोद समायना नंगौ। |
| Bangalore is famous for software | बांगालराव software थाखाय मुंदांखा। |
| Guwahati is the main city of Assam | गुवाहाटीया आसामनि गुबै नोगोर। |
| Chandni Chowk is located opposite the Red Fort | चान्दनि चौका रेड फर्टनि उलथा फारसे दं। |

technique is already built in Moses. The following command is used to find the BLEU score in the proposed system:

```
~/mosesdecoder/scripts/generic/multi-bleu.perl –lc    ~/corpus/training/tourism.eng-
bod.true.bd    < ~/working/output.tourism.eng-bod.bd
```

BLEU score 65.09 is recorded for the proposed system. It has been observed that more the number of sentences of each language to train the system, more the improvement in the BLEU score. A higher BLEU score denotes better translation.


# 6    Conclusion

Machine Translation is a very helpful system for us to translate the huge amount of text from a natural language to another language using computers. Primarily the proposed MT system is implemented for high quality and accurate translation result from English to Bodo language without any human interaction. In the proposed system, Phrase-Based SMT approach, Moses, N-gram technique, BLEU technique and tourism domain English to Bodo parallel corpora are used for implementation. The proposed system is tested with different numbers of tourism domain parallel sentences of English and Bodo languages and achieved different translation results. From the proposed system, it has been observed that the Phrase-Based SMT approach using a large amount of parallel corpora for good and accurate translation results. Computerized information for Bodo language is very scarce and no MT system for Bodo language is available till date. Hence, it is expected that the proposed SMT system would be immensely helpful for Bodo people as well as for others in India and abroad.

## 7 Scope for Future Research

The following possible tasks and ideas can be added to extend the proposed research work.

1. Adding greater number of sentences in the English to Bodo parallel corpora for better translation results.
2. Adding transliteration in the proposed system to improve the quality of translation.
3. Putting the proposed system on the web-based portal.
4. Design and develop English⇔Bodo Phrase-Based SMT system.
5. Design the English to Bodo SMT system to support in Mobile Phone application.

## References

1. Bhattacharyya, P.: Machine translation. IIT Bombay, Mumbai (2001)
2. Antony, P.J.: Machine translation approaches and survey for Indian languages. Comput. Linguist. Chin. Lang. Process. **18**(1), 47–78 (2013)
3. https://en.wikipedia.org/wiki/Bodo_language
4. https://en.wikipedia.org/wiki/English_language
5. Kathiravan, P., Makila, S., Prasanna, H., Vimala, P.: Over view—The machine translation in NLP. Int. J. Sci. Adv. Technol. **2**(7), 19–25 (2016)
6. Hutchins, W.J.: Machine translation: History of research and applications. University of East Anglia, UK (1995)
7. Brunning, J.: Alignment models and algorithms for statistical machine translation (Thesis). Cambridge University, UK (2010)
8. Koehn, P.: Statistical machine translation. Cambridge University Press, New York (2009)
9. Islam, M.Z., Eisele, A.: English to Bangla phrase based statistical machine translation (Thesis). Germany University, Saarland (2009)
10. Sharma, N., Bhatia, P.: English to Hindi statistical machine translation (Thesis). Thapar University, Patiala (2011)
11. Bharati, A., Moona, R., Reddy, P., Sankar, B., Sharma, D.M., et al.: Machine translation: the Shakti approach. In: Proceeding of the 19th International Conference on Natural Language Processing, pp. 1–7, Dec 2003. MT-Archive, India (2003)
12. Godase, A., Govilkar, S.: Machine translation development for Indian languages and its approaches. Int. J. Natl. Lang. Comput. (IJNLC) **4**(2), 55–74 (2015)
13. Das, P., Baruah, K.K.: Assamese to English statistical machine translation integrated with a transliteration module. Int. J. Comput. Appl. **100**(5), 20–24 (2014)
14. Singh, M.T., Borgohain, R., Gohain, S.: An English-assamese machine translation system. Int. J. Comput. Appl. **93**(4), 01–06 (2014)
15. George, A.: English To Malayalam statistical machine translation system. Int. J. Eng. Res. Technol. (IJERT) **2**(7), 640–647 (2013)
16. Koehn, P.: MOSES (User Manual and Code Guide). Statistical machine translation system. University of Edinburgh, UK (2014)
17. Uszkoreit, H.: Survey of machine translation evaluation. EuroMatrix Project, Germany (2007)

# Part VII
# Advanced Computing: Case Study

# Modelling Sustainable Procurement: A Case of Indian Manufacturing Firm

**Harpreet Kaur and Surya Prakash Singh**

**Abstract** In today's dynamic and challenging environment, business decisions are driven not only by cost but also by environmental legislations. In manufacturing industries, procurement is a crucial operation responsible for cost, quality and delivery of final product. Unlike purchasing, procurement involves strategic planning of lot-sizes to be ordered from suppliers and carriers. The paper proposes a sustainable procurement model (SPM) integrating lot-sizing, supplier and carrier selection in a cap-and-trade environment, optimizing cost and carbon emissions. The model is illustrated with the help of a case study of an Indian manufacturing firm procuring raw materials from multiple suppliers using multiple carriers.

**Keywords** Integrated procurement · Sustainability · Carbon emissions · Lot-sizing · Supplier selection · Carrier selection · Modelling

## 1 Introduction

The role of procurement has become highly significant in today's dynamic business scenario, especially in manufacturing industries where procurement costs are substantial. To meet the demands timely, an effective and efficient procurement model is required. As these firms are dependent on suppliers and carriers for procurement, a judicious selection of suppliers and carriers is very important. There is much work done on qualitative and quantitative selection of suppliers which can be found at Ware et al. [1], Aggarwal and Singh [2], Kaur et al. [3]. Moreover, procurement is a multi-period activity, therefore, lot-sizing of procured parts also plays an essential role. In this context, a number of models integrated with supplier selection in

H. Kaur · S.P. Singh (✉)
Department of Management Studies, Indian Institute of Technology Delhi,
New Delhi, India
e-mail: surya.singh@gmail.com

H. Kaur
e-mail: harpreetk.iitd@gmail.com

© Springer Nature Singapore Pte Ltd. 2018
R.K. Choudhary et al. (eds.), *Advanced Computing
and Communication Technologies*, Advances in Intelligent Systems
and Computing 562, https://doi.org/10.1007/978-981-10-4603-2_21

procurement are proposed in literature, e.g. Tempelmeier [4], Basnet and Leung [5], Aissaoui et al. [6], Ustun and Demı [7], Rezaei and Davoodi [8]; however, these models do not take into account the carbon cost.

The manufacturing industries these days are enforced with legislations to reduce the carbon emissions [9, 10]. Recently, some work has been done on integrating carbon emission constraints in procurement lot-sizing and supplier selection [11–20]. However, these models lack logistics aspects. It has been observed that there is a huge amount of logistics involved in procurement process, therefore, the type of carrier and its capacity contributes significantly a lot in emissions caused during procurement. In this context, Liao and Rittscher [21] proposed an integrated model for order allocation, supplier and carrier selection. However, the proposed model is limited to single period and does not consider carbon emissions. Similarly, Songhori et al. [22] proposed integrated model jointly considering lot-sizing, supplier and carrier selection for single item. Recently, Kaur and Singh [23] proposed a multi-part procurement problem with integrated lot-sizing, supplier and carrier selection, but the proposed model (ibid) does not consider carbon emissions. Therefore, the paper proposes an Integrated Sustainable Procurement Model (ISPM), integrating lot-sizing, supplier and carrier selection for multi-period, multi-part, multi-supplier and multi-carrier problem in the presence of carbon emissions using the prevalent carbon cap-and-trade policy. Carbon cap, a firm can generate during the procurement, is thus provided over the total emissions.

Following a brief introduction, the paper is organized as follows: Sect. 2 presents the proposed sustainable procurement model. Section 3 discusses the case of the manufacturing industry followed by results and discussions. Section 5 presents conclusions.

## 2 Sustainable Procurement Model

The proposed procurement model addresses a multi-period, multi-part, multi-supplier and multi-carrier problem. The model minimizes the total procurement cost including raw material cost, ordering cost, transportation cost, holding cost and carbon emissions cost subjected to supplier capacity, carrier capacity, warehouse capacity and lead-time constraints. The model assumes demand, supplier capacity and carrier capacity as known and constant. The carbon emissions from ordering, transporting and holding of parts in the model are assumed to be linear. The list of notations used to model are shown in Table 1.

$$\text{Minimize } Z = Z_1 + Z_2 + Z_3 + Z_4 + Z_5 \tag{1}$$

$$Z_1 = \sum_t \sum_i \sum_j \sum_m p_{tij} x_{tijm} \tag{1a}$$

**Table 1** List of notations for sustainable procurement mode

| Notations | Description |
|---|---|
| $t$ | Index for time periods |
| $i$ | Index for parts |
| $j$ | Index for suppliers |
| $m$ | Index for carriers |
| $x_{tijm}$ | Order allocation in $t$th period of $i$th part procured through $j$th supplier from $m$th carrier |
| $U_{tijm}$ | 1 if in $t$th period the $i$th part is through $j$th supplier from $m$th carrier, Else 0 |
| $Y$ | Carbon emissions sold or bought across entire period |
| $I_{ti}$ | $i$th part inventory carried from $t$th period to $t^{(t+1)}$th period |
| $D_{ti}$ | Demand in $t$th period for $i$th part |
| $P_{tij}$ | Cost of purchasing in $t$th period of $i$th part from $j$th supplier |
| $t_{tjm}$ | Cost of transportation in $t$th period through $j$th supplier from $m$th carrier |
| $O_{ti}$ | Cost of ordering in $t$th period of $i$th part |
| $h_{ti}$ | Cost of holding inventory in $t$th period for $i$th part |
| $C_{tij}$ | Capacity in $t$th period of $j$th supplier for $i$th part |
| $\Omega_{jm}$ | Available truck load capacity of $m$th carrier with $j$th supplier |
| $V_{tjm}$ | Total number of $m$th carriers available in $t$th period with $j$th supplier |
| $\alpha$ | Carbon emissions quota (in tonnes) for entire planning horizon |
| $C$ | Carbon price per unit (tonne) |
| $F_{tm}, F_{tom}$ | Carbon emission for $x_{tijm}$. $F_{mt}$ is fixed carbon emissions for empty carrier. $F_{omt}$ is the variable carbon emission |
| $E_{ot}$ | Carbon emissions caused for placing an order in each time period |
| $E_{tw}$ | Carbon emissions due to holding a part unit in each time period |
| $UL_{ti}$ | Upper tolerance of lead time in $t$th period for $i$th part |
| $LL_{ti}$ | Lower tolerance of lead time for $t$th period for $i$th part |
| $L_{tjm}$ | Lead time in $t$th period through $j$th supplier from $m$th carrier |
| $dj$ | Distance (km) of $j$th supplier from the buyer |
| $mil_m$ | Mileage (km/l) of $m$th carrier |

$$Z_2 = \sum_t \sum_i \sum_j O_{ti} U_{tijm} \tag{1b}$$

$$Z_3 = \sum_t \sum_i \sum_j \sum_m t_{tjm} x_{tijm} \tag{1c}$$

$$Z_4 = \sum_t \sum_i h_{ti} I_{ti} \tag{1d}$$

$$Z_5 = C * Y \tag{1e}$$

$$I_{t-1\,i} + \sum_{j}\sum_{m} x_{tijm} - I_{ti} = D_{ti} \quad \forall t,i \tag{2}$$

$$x_{tijm} \leq \left(\sum_{t} D_{ti}\right) U_{tijm} \quad \forall t,i,j,m \tag{3}$$

$$\sum_{m} x_{tijm} \leq C_{tij} \quad \forall t,i,j \tag{4}$$

$$\sum_{i} x_{tijm} \leq \Omega_{jm} V_{tjm} \quad \forall t,j,m \tag{5}$$

$$\sum_{t}\sum_{i}\sum_{j}\sum_{m} (F_{tm} + F_{tom}) U_{ijmt} + \sum_{t}\sum_{i}\sum_{j}\sum_{m} E_{to} U_{tijm}$$
$$+ \sum_{t}\sum_{i} E_{tw} I_{ti} = \alpha + Y \tag{6}$$

$$F_{tom} = \frac{d_j}{mil_m} * \text{emission factor} * x_{tijm} \quad \forall t,i,j,m \tag{7}$$

$$LL_{ti} \leq U_{tijm} l_{tjm} \leq UL_{ti} \quad \forall t,i,j,m \tag{8}$$

$$x_{tijm} \text{ and } I_{ti} \geq 0 \text{ and integer} \tag{9}$$

$$U_{tijm} \in \{0,1\} \tag{10}$$

$$Y \text{ is unstricted sign} \tag{11}$$

Equation (1) minimizes procurement cost comprising of raw material cost (1a), ordering cost (1b), transportation cost (1c), holding cost (1d) and, carbon emissions cost (1e). Equation (2) balances the inventory for each period. Equation (3) prevents excess procurement of parts. Equations (4) and (5) take care of supplier and carrier capacity, respectively. Equation (6) balances the total carbon emissions caused due to ordering, holding and transporting parts with total allowable emissions and additional emissions can be bought or sold. Equation (7) relates variable transportation emissions as a function of distance travelled, mileage of carrier and load carried by the carrier. Equation (8) is the lead-time constraint to ensure timely delivery of parts. The integer and non-negative value of parts ($x_{ijmt}$) and inventory ($I_{it}$) is ensured by Eq. (9). Equation (10) puts binary restriction on supplier and carrier selection. Equation (11) represents unrestricted sign for additional emissions bought or sold.

# 3  Case Study

Kapsons Industries Private Limited is a manufacturing firm having multiple divisions operating separately in Jalandhar, Punjab and Pune, Maharashtra. The case study pertains to the motor division located in Jalandhar, Punjab. The firm accepts orders (different models of motors) from various national and international partners. To deliver the ordered quantities, the firm in-turn orders coil sheets of various grades from multiple national and international suppliers. The firm requires an optimal procurement plan for multi-periods, multi-parts, multi-suppliers and multi-carriers. Firm serves its global buyers where sustainability is a key issue. Therefore, the firm attempts to optimize its total procurement cost and carbon emission cost through modelling. The data used in the model is collected from the secondary sources. However, data not available is derived using standard parameters available with the firm. Following are procedures adopted to calculate data from secondary sources of the firm.

1. The demand of various parts is calculated using different utility factors for different products and is measured in kgs.
2. Raw material cost includes the transportation from supplier to supplier's warehouse.
3. The transportation cost is incurred by the firm itself for transportation from supplier's warehouse to the firm.
4. Excess inventory is held at supplier's warehouse and rent is paid by the firm.
5. The firm uses two carrier types of capacity 15 and 20 tonne.
6. The suppliers offer different capacities and lead times (refer Appendix I).
7. Carbon emissions data is not recorded and therefore calculated using standard parameters.

The supplier information is provided in Table 2. Based on secondary data, the problem is modelled (as discussed in Sect. 2) and solved in LINGO 10. Results are discussed in Sect. 4.

**Table 2**  Supplier information

| | Supplier | Warehouse | Distance from firm (km) |
|---|---|---|---|
| $S_1$ | SAIL | Ludhiana | 80 |
| $S_2$ | POSCO | Gurgaon | 460 |
| $S_3$ | Thyssen | Bengaluru | 2700 |
| $S_4$ | JSW | Ludhiana | 80 |
| $S_5$ | Marathon elec | Faridabad | 410 |

# 4  Case Results and Discussions

The procurement plan for three months for the case discussed is given in Table 3. The orders allocated to different suppliers and carriers for all the parts in all the periods can also be observed from Table 3. The overall cost incurred in

**Table 3**  Optimal procurement plan for motor division unit at Jalandhar, Punjab

|   |   | $T_1$ | $T_2$ | $T_3$ |
|---|---|---|---|---|
| $P_1$ | Demand | 166458.7 | 168123.3 | 164794.1 |
|  | Lot-sizing | 269207.1 | 72374.89 | 157794.1 |
|  | Order allocation | $X_{(1, 1, 4, 1)} = 7000$; $X_{(1, 1, 4, 2)} = 166458.7$; $X_{(2, 1, 1, 2)} = 108123.3$; $X_{(2, 1, 4, 2)} = 60,000$; $X_{(3, 1, 1, 2)} = 97794.11$; $X_{(3, 1, 4, 2)} = 60,000$ | | |
| $P_2$ | Demand | 1,014,016 | 1,024,156 | 1,003,876 |
|  | Lot-sizing | 1,021,016 | 1,024,156 | 996,876 |
|  | Order allocation | $X_{(1, 2, 1, 2)} = 121016.2$; $X_{(1, 2, 3, 1)} = 107,000$; $X_{(1, 2, 3, 2)} = 793,000$; $X_{(2, 2, 1, 2)} = 110449.6$; $X_{(2, 2, 2, 2)} = 13706.76$; $X_{(2, 2, 3, 1)} = 100,000$; $X_{(2, 2, 3, 2)} = 800,000$; $X_{(3, 2, 1, 2)} = 96876.01$; $X_{(3, 2, 3, 1)} = 100,000$; $X_{(3, 2, 3, 2)} = 800,000$ | | |
| $P_3$ | Demand | 14,800 | 14,948 | 14,652 |
|  | Lot-sizing | 21,800 | 14,948 | 7652 |
|  | Order allocation | $X_{(1, 3, 4, 1)} = 7000$; $X_{(1, 3, 4, 2)} = 14,800$; $X_{(2, 3, 4, 2)} = 14,948$; $X_{(3, 3, 4, 2)} = 7652$ | | |
| $P_4$ | Demand | 21,000 | 21,210 | 20,790 |
|  | Lot-sizing | 28,000 | 21,210 | 13,790 |
|  | Order allocation | $X_{(1, 4, 3, 1)} = 21,000$; $X_{(1, 4, 3, 2)} = 7000$; $X_{(2, 4, 3, 1)} = 21,210$; $X_{(3, 4, 3, 1)} = 13,790$ | | |
| $P_5$ | Demand | 560863.7 | 566472.4 | 555255.1 |
|  | Lot-sizing | 567863.7 | 566472.4 | 548255.1 |
|  | Order allocation | $X_{(1, 5, 1, 2)} = 560863.7$; $X_{(1, 5, 2, 2)} = 7000$; $X_{(2, 5, 1, 2)} = 566472.4$; $X_{(3, 5, 1, 2)} = 548255.1$ | | |
| $P_6$ | Demand | 28,500 | 28,785 | 28,215 |
|  | Lot-sizing | 35,500 | 28,785 | 21,215 |
|  | Order allocation | $X_{(1, 6, 2, 2)} = 7000$; $X_{(1, 6, 4, 2)} = 28,500$; $X_{(2, 6, 4, 2)} = 28,785$; $X_{(3, 6, 4, 2)} = 21,215$ | | |
| $P_7$ | Demand | 4375 | 4418.75 | 4331.25 |
|  | Lot-sizing | 11,375 | 4418.75 | – |
|  | Order allocation | $X_{(1, 7, 1, 2)} = 4375$; $X_{(1, 7, 2, 2)} = 4375$; $X_{(1, 7, 4, 2)} = 2625$; $X_{(2, 7, 2, 2)} = 4418.75$ | | |
| $P_8$ | Demand | 14806.7 | 14954.77 | 14658.63 |
|  | Lot-sizing | 21806.7 | 14954.77 | 7658.633 |
|  | Order allocation | $X_{(1, 8, 1, 1)} = 7000$; $X_{(1, 8, 1, 2)} = 14806.7$; $X_{(2, 8, 1, 2)} = 14954.77$; $X_{(3, 8, 1, 2)} = 7658.633$ | | |

(continued)

**Table 3** (continued)

|  |  | $T_1$ | $T_2$ | $T_3$ |
|---|---|---|---|---|
| $P_9$ | Demand | 12217.06 | 12339.23 | 12094.89 |
|  | Lot-sizing | 19217.06 | 12339.23 | 5094.889 |
|  | Order allocation | $X_{(1, 9, 1, 2)} = 7000; X_{(1, 9, 4, 2)} = 12217.06; X_{(2, 9, 4, 2)} = 12339.23; X_{(3, 9, 4, 2)} = 5094.889$ | | |
| $P_{10}$ | Demand | 39941.24 | 40340.65 | 39541.83 |
|  | Lot-sizing | 46941.24 | 40340.65 | 32541.83 |
|  | Order allocation | $X_{(1, 10, 1, 2)} = 7000; X_{(1, 10, 2, 2)} = 39941.24; X_{(2, 10, 2, 2)} = 40340.65; X_{(3, 10, 2, 2)} = 32541.83$ | | |

procurement for three months is found to be INR 414701282. The raw material cost is INR 399490299, ordering cost is INR 53600, transportation cost is INR 13859220, Holding cost is INR 582006 and carbon emission cost is INR 716157.

## 5   Conclusions

The paper proposes an integrated sustainable procurement model (ISPM) for optimal lot-sizing, supplier and carrier selection for a multi-period, multi-part, multi-supplier and multi-carrier problem. The model optimizes the total procurement cost and the carbon emissions cost too. The model is further validated by analyzing the case of a manufacturing firm located at Jalandhar, Punjab, India by providing an optimal order to be procured from respective supplier using respective carrier such that total procurement cost and carbon emissions cost are simultaneously minimized.

## Appendix I

Secondary data for manufacturing firm located at Jalandhar, Punjab

| Parts | $OC_{ti}$ | $D_{ti}$ (kg)[a] | $PC_{tij}^{b}$ | $SC_{tij}$(*10,000) (kg)[b] | Suppliers | $I_{tjm}^{c}$ | $TC_{tim}^{c}$ |
|---|---|---|---|---|---|---|---|
| $P_1$(40.5) | 1000 | 166458.7, 168123.3, 164794.113 | 68, 74.9, 73.3, 67.2, 85.1 | 36, 18, 12, 60, 9 | $S_1$ | 1, 2 | 5, 3 |
| $P_2$(41.2) | 1000 | 1,014,016, 1,024,156, 1003876.008 | 69.63, 70.86, 67.6, 76.6, 88.6 | 180, 120, 90, 30, 60 | $S_2$ | 3, 4 | 4, 2 |
| $P_3$(41.5) | 1200 | 14,800, 14,948, 14,652 | 68, 74.9, 73.3, 67.2, 85.1 | 36, 18, 20, 6, 9 | $S_3$ | 4, 5 | 4, 2 |

(continued)

(continued)

| Parts | $OC_{ti}$ | $D_{ti}$ (kg)[a] | $PC_{tij}^b$ | $SC_{tij}$(*10,000) (kg)[b] | Suppliers | $l_{tjm}^c$ | $TC_{tim}^c$ |
|---|---|---|---|---|---|---|---|
| $P_4$(42.2) | 1200 | 21,000, 21,210, 20,790 | 69.63, 70.86, 67.6, 76.6, 88.6 | 180, 120, 90, 30, 180 | $S_4$ | 1, 1 | 3, 1 |
| $P_5$(43.9) | 1400 | 560863.7, 566472.4, 555255.1 | 73.8, 75.5, 76.39, 78.51, 83.41 | 1110, 60, 45, 60, 30 | $S_5$ | 1, 2 | 7, 4 |
| $P_6$(45.5) | 1200 | 28,500, 28,785, 28,215 | 82.8, 83.6, 86.45, 84.5, 85 | 36, 18, 120, 6, 9 | Carbon calculation | | Others |
| $P_7$(45.6) | 1200 | 4375, 4418.75, 4331.25 | 79, 79.8, 82, 81, 80 | 180, 120, 90, 30, 60 | $F_{tm}$ | 0.15, 0.2 | $ul_{ti} = 7$ |
| $P_8$(50) | 1000 | 14806.7, 14954.77, 14658.633 | 85, 92.5, 87.5, 91, 92 | 36, 18, 12, 6, 9 | $E_{to}$ | 0.01 | $ll_{ti} = 0$ |
| $P_9$(54) | 1200 | 12217.06, 12339.23, 12094.8894 | 95, 96, 99.36, 96, 96 | 180, 120, 90, 30, 60 | $E_{tw}$ | 0.003 | $cc_{Jm} = 15$, 20 (mtn) |
| $P_{10}$(58) | 1000 | 39941.24, 40340.65, 39541.8276 | 101.5, 102, 103.82, 106, 118.9 | 9, 36, 18, 12, 6 | $Mil_m$ | 14, 12 | $V = 20, 30$ |

[a]Means that data belongs to set $\{T_1, T_2, T_3\}$
[b]Means that data belongs to set $\{S_1, S_2, S_3, S_4, S_5\}$
[c]Means that data belongs to set $\{M_1, M_2\}$

# References

1. Ware, N.R., Singh, S.P., Banwet, D.K.: A mixed-integer non-linear program to model dynamic supplier selection problem. Expert Syst. Appl. **41**(2), 671–678 (2014)
2. Aggarwal, R., Singh, S.P.: Chance constraint-based multi-objective stochastic model for supplier selection. Int. J. Adv. Manuf. Technol. **79**(9–12), 1707–1719 (2015)
3. Kaur, H., Singh, S.P., Glardon, R.: An integer linear program for integrated supplier selection: a sustainable flexible framework. Glob. J. Flex. Syst. Manag. **17**(2), 1–22 (2016)
4. Tempelmeier, H.: A simple heuristic for dynamic order sizing and supplier selection with time-varying data. Prod. Oper. Manag. **11**(4), 499 (2002)
5. Basnet, C., Leung, J.M.: Inventory lot-sizing with supplier selection. Comput. Oper. Res. **32**(1), 1–14 (2005)
6. Aissaoui, N., Haouari, M., Hassini, E.: Supplier selection and order lot sizing modeling: a review. Comp. Oper. Res. **34**(12), 3516–3540 (2007)
7. Ustun, O., Demı, E.A.: An integrated multi-objective decision-making process for multi-period lot-sizing with supplier selection. Omega **36**(4), 509–521 (2008)
8. Rezaei, J., Davoodi, M.: Multi-objective models for lot-sizing with supplier selection. Int. J. Prod. Econ. **130**(1), 77–86 (2011)
9. Simpson, D., Power, D.: Use the supply relationship to develop lean and green suppliers. Supply Chain Manag. Int. J. **10**(1), 60–68 (2005)

10. Geffen, C., Rothenberg, S.: Suppliers and environmental innovation. Int. J. Oper. Prod. Manag. **20**(2), 166–186 (2000)
11. Ciliberti, F., Pontrandolfo, P., Scozzi, B.: Logistics social responsibility: standard adoption and practices in Italian companies. Int. J. Prod. Econ. **113**, 88–106 (2008)
12. Bai, C., Sarkis, J.: Integrating sustainability into supplier selection with grey system and rough set methodologies. Int. J. Prod. Econ. **124**, 252–264 (2010)
13. Kumar, A., Jain, V.: Supplier selection: a green approach with carbon footprint monitoring. In: 8th International Conference on Supply Chain Management and Information Systems (SCMIS), Hong Kong, pp. 1–8 (2010)
14. Seuring, S., Muller, M.: From a literature review to a conceptual framework for sustainable supply chain management. J. Clean. Prod. **16**(15), 1699–1710 (2008)
15. Benjaafar, S., Li, Y., Daskin, M.: Carbon footprint and the management of supply chains: insights from simple models. IEEE Trans. Autom. Sci. Eng. **10**(1), 99–116 (2013)
16. Helmrich, M.R., Van den Heuvel, W., Wagelm, A.P.: The economic lot-sizing problem with an emission constraint. 2nd International Workshop on Lot Sizing, Istanbul, Turkey, pp. 45–48 (2011)
17. Battini, D., Persona, A., Sgarbossa, F.: A sustainable EOQ model: theoretical formulation and applications. Research report, Department of Management and Engineering, University of Padua (2012)
18. Presley, A., Meade, L., Sarkis, J.: A strategic sustainability justification methodology for organizational decisions: a reverse logistics illustration. Int. J. Prod. Res. **45**(18–19), 4595–4620 (2007)
19. Tao, Z., Guiffrida, A.L., Troutt, M.D.: A green cost based economic production/order quantity model. Proceedings of the 1st Annual Kent State International Symposium on Green Supply Chains, Canton, Ohio, US (2010)
20. Hsu, C.W., Chen, S.H., Chiou, C.Y.: A model for carbon management of supplier selection in green supply chain management. IEEE International Conference on Industrial Engineering and Engineering Management (IEEM), Singapore, pp. 1247–1250 (2011)
21. Liao, Z., Rittscher, J.: Integration of supplier selection, procurement lot sizing and carrier selection under dynamic demand conditions. Int. J. Prod. Econ. **107**(2), 502–510 (2007)
22. Songhori, M.J., Tavana, M., Azadeh, A., Khakbaz, M.Z.: A supplier selection and order allocation model with multiple transportation alternatives. Int. J. Adv. Manuf. Technol. **52**, 365–376 (2011)
23. Kaur, H., Singh, S.P.: Modelling flexible procurement problem. In: Sushil, Bhal, K.T., Singh, S.P. (eds.) Managing Flexibility: People, Process, Technology and Business, pp. 147–170. Springer, New Delhi (2016)

# Part VIII
# Advanced Computing: Miscellaneous—Social Networks

# Key Author Analysis in Research Professionals' Collaboration Network Based on Collaborative Index

Anand Bihari and Sudhakar Tripathi

**Abstract** Generally, research work is done by the group of researchers and they form a network called Research Professionals' Collaboration Network, which acts as a social network. In social Network, the key nodes are defined by the centrality measures, while the key researchers in the research community are defined by the h-index. In this paper, the collaborative index, a combination of social network characteristics and the citation based index h-index has been discussed. The lobby-index, weighted network lobby-index, h-degree and c-index have been used to discover the key researchers in the community.

**Keywords** Social networks · Research collaboration · Lobby-index · Wl-index · C-index · h-degree

## 1 Introduction

During the last decade, it seems that the most of the research projects are too large to complete by an individual researcher or a group of researchers from same subject area. So the collaboration amongst researchers span over multiple subject areas. The analysis of collaboration network with respect to social network analysis is a good measure [1] to evaluate the scientific impact of the researcher. But the evaluation of research work is based on the citation count. H-index is a good measure to evaluate the scientific impact of individuals, when we considered citation count of articles. But most of the research works are completed by the group of researchers and it requires collaboration with other researcher who may be from same or different areas. The evaluation of scientific impact of individual will require a hybrid

A. Bihari (✉) · S. Tripathi
Department of Computer Science & Engineering,
National Institute of Technology Patna, Patna, Bihar, India
e-mail: anand.cse15@nitp.ac.in

S. Tripathi
e-mail: stripathi.cse@nitp.ac.in

© Springer Nature Singapore Pte Ltd. 2018
R.K. Choudhary et al. (eds.), *Advanced Computing
and Communication Technologies*, Advances in Intelligent Systems
and Computing 562, https://doi.org/10.1007/978-981-10-4603-2_22

approach of h-index and social network analysis metrics. In this paper construction of the collaboration network is discussed followed by discussions on different types of collaborative indexes viz. combination of h-index and centrality measures of social network analysis such as lobby-index [2], weighted network lobby-index [3] and h-degree [3] for evaluating the scientific impact of the individuals. To validate the analysis of collaborative index, an experimental analysis has been made.

## 2 Background

Newman [4] recommended the idea of weighted collaboration network based on the quantity of co-authors. The co-authorship weight between two collaborators is the total number of publication, which is published together and their ratio. Then, they use power law to estimate the overall co-authorship weight among researchers. Abbasi et al. [5, 6] built a weighted co-authorship network and used social network analysis metrics for evaluation of performance of individual researchers. Here weight is the total number of collaboration between researchers. Bihari et al. [7] discussed the undirected weighted collaboration network of researcher and discovered the prominent researcher based on centrality measures (degree, closeness, betweenness and eigenvector centrality) of social network and citation-based indicators (Frequency, Citation count, h-index, g-index and i10-index). In this article author used total citation count earn by the collaborators. Arnaboldi et al. [8] deliberated the egocentric network of researchers and studied the performance of all researchers. In this paper, h-index and the average amount of authors have been used to evaluate the scientific influence of researchers and furthermore discover the relation among h-index and average number of authors. Bihari et al. [9] used the maximum spanning tree to remove weak edges from a collaborative network of researchers and used social network centrality measures to appraise the scientific impact of individual and find out the prominent actor in the network of a given dataset.

## 3 Methodology

### 3.1 Lobby-Index (l-Index)

It is used to evaluate the node importance in the undirected non-weighted network. The lobby-index (l-index) [2] of a node $k$ is the largest integer $n$ such that the node $k$ has at least $n$ direct neighbours with at least $n$ degree and the rest of the nodes may have $n$ or less. The main objective of this index is to evaluate the node in a network is based on their neighbours' strength. Basically, it is inspired by the h-index [10] and has combination with the degree in the aspects of social network [1].

## 3.2    Weighted Network Lobby-Index (Wl-Index)

It is used to evaluate the node importance in weighted undirected network. The Weighted network lobby-index [3] of node $k$ is the highest value $n$ such that the node $k$ has at least $n$ neighbours which have at least $n$ node strength. Here the node strength is the sum of the collaboration weight with the neighbour nodes.

## 3.3    h-Degree

It is used to evaluate the node in the weighted network based on their neighbours' strength as well as edge strength. The h-degree [3] of a node $k$ is the largest rank $h$ such that the node $k$ has at least $h$ neighbours and the strength of the edge between $k$ and neighbour node is at least $h$. It is the combination of h-index and the degree of a node.

## 3.4    Collaborative Index (C-Index)

The collaborative index [11] of node $k$ is largest integer $m$ such that the node $k$ has at least $m$ neighbours which has individually $m$ productive strength and the rest of the neighbours has $m$ or less neighbours with any productive strength each.

## 4    Data Collection and Analysis

For analysis of key author in the collaboration network, it will require collaboration data like total number of articles or total number of citation count of collaborator. The collaboration data were downloaded from IEEE Xplore for the period of Jan-2000 to July-2014 which was used in [7, 9, 12, 13] including journal and conference proceedings. The raw data contains so many fields, but it requires only article name, authors, article type and citation count. The raw data set requires cleaning for removing unambiguous data. After cleaning of published data, 26,802 articles and 61,546 authors were available for analysis. In the dataset, the two papers were composed of the 100 or more than 100 authors, i.e. 123 and 119, three papers were composed by more than 50 authors, 36 papers were composed by 20 or more than 20 authors, 437 paper were composed by the 10 or more than 10 authors, 5167 papers were written by 5 or more authors, 4526, 6772, 7036, 3236 papers are composed by exact 4, 3, 2, 1 author(S), respectively. After analysis of collaborative data, we can say that most of the article was published with the collaboration of 2–4 authors.

Author Lau, Y.Y. published 54 articles in different journal and conferences. Author Sasaki, M. published works with 63 different authors, around 17% authors work individually and 38% authors have collaborated with 2 authors.

## 5  Collaboration Network of Research Professionals' an Example

The construction of collaboration network has been required to evaluate the scientific impact of individual researcher. The following techniques are used to collaborate network construction, author collaboration extraction and weight calculation (based on total publication count, citation count and collaboration age) [14, 15]. For example, we considered three articles: Art1, Art2 and Art3 details are shown in Table 1.

Here the collaboration between R and S is higher than others and collaboration weight based on total publication count, citation count and collaboration age are 2, 110 and 3, respectively. The network is constructed by using python and networkX [16]. The network is like Fig. 1.

| Table 1 Example of publication and their authors | Paper ID | Author's name | Citation count | Publication year |
|---|---|---|---|---|
| | ART1 | X, N, R, S | 10 | 2008 |
| | ART2 | P, S, Y | 22 | 2010 |
| | ART3 | M, R, S | 100 | 2011 |

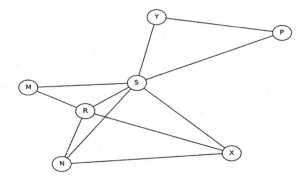

**Fig. 1** Research professionals' collaboration network an example analysis and result

# 6    Analysis and Result

Three weighted undirected networks have been constructed for evaluation of scientific impact of individual. The weight is like total publication count, total citation
count and the total collaboration age and convey the lobby-index, wl-index, h-degree
and c-index. Based on computational results the lobby-index of authors are equal in
different weighted network because weight does not consider in this index computation. But wl-index, h-degree and c-index have variation due to their weight
changes. The variation in result in different weighted network are shown in Figs. 2, 3
and 4 of wl-index, h-degree and c-index, respectively. In collaboration network,
some of the collaboration has high collaboration age, but their productivity is less or

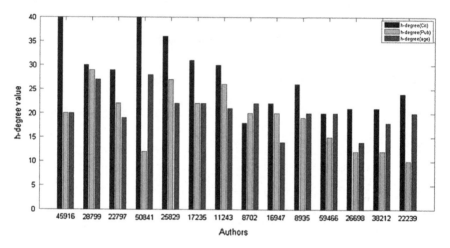

**Fig. 2**  Top 14 authors wl-index in all three-weighted collaboration network

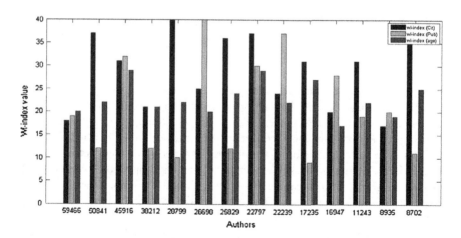

**Fig. 3**  Top 14 authors h-degree in all three-weighted collaboration network

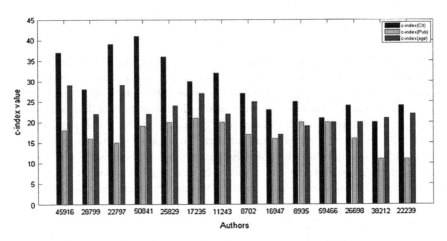

**Fig. 4** Top 14 authors c-index in all three-weighted collaboration network

**Table 2** Collaboration index result in citation count weighted network

| Sl. No. | Name of author | l-index | wl-index | h-degree | c-index |
|---------|----------------|---------|----------|----------|---------|
| 1 | Osterwinter, Heinz | 42 | 31 | 31 | 37 |
| 2 | Hasegawa, T. | 42 | 40 | 40 | 28 |
| 3 | Shimizu, K. | 42 | 37 | 37 | 39 |
| 4 | Yoshino, K. | 42 | 37 | 37 | 41 |
| 5 | Fiore, A. | 36 | 36 | 36 | 36 |
| 6 | Tomkos, I. | 32 | 31 | 31 | 30 |
| 7 | Klonidis, D. | 32 | 31 | 31 | 32 |
| 8 | Kouloumentas, C. | 32 | 37 | 37 | 27 |
| 9 | Lau, Y.Y. | 27 | 20 | 20 | 23 |
| 10 | Gilgenbach, R.M. | 27 | 17 | 17 | 25 |
| 11 | Freude, W. | 25 | 18 | 18 | 21 |
| 12 | Koos, C. | 25 | 25 | 25 | 24 |
| 13 | Leuthold, J. | 25 | 21 | 21 | 20 |
| 14 | Becker, J. | 25 | 24 | 24 | 24 |

may have less impact compare to other collaborators who have less collaboration age. In another case, some collaboration produces a huge amount of publication, but their impact is very less than others who produced less amount of publication with high impact. For computation of collaborative index, we used Python and NetworkX [16] as a computing platform. In all three different weighted network author **Osterwinter, Heinz** is the key author in the network based on lobby-index. In citation weighted network, author **Hasegawa, T. and Yoshino, K.** are key author based on Wl-index while author **Osterwinter, Heinz and Yoshino, K.** are a key author based on h-degree and c-degree, respectively. The collaborative index result for citation count weighted collaboration network shown in Table 2.

**Table 3** Collaboration index result in publication count weighted network

| Sl. No. | Name of author | l-index | wl-index | h-degree | c-index |
|---------|----------------|---------|----------|----------|---------|
| 1 | Osterwinter, Heinz | 42 | 32 | 20 | 18 |
| 2 | Hasegawa, T. | 42 | 40 | 29 | 16 |
| 3 | Shimizu, K. | 42 | 37 | 22 | 15 |
| 4 | Yoshino, K. | 42 | 12 | 12 | 19 |
| 5 | Fiore, A. | 36 | 30 | 27 | 20 |
| 6 | Tomkos, I. | 32 | 28 | 22 | 21 |
| 7 | Klonidis, D. | 32 | 20 | 26 | 20 |
| 8 | Kouloumentas, C. | 32 | 27 | 20 | 17 |
| 9 | Lau, Y.Y. | 27 | 19 | 20 | 16 |
| 10 | Gilgenbach, R.M. | 27 | 11 | 19 | 20 |
| 11 | Freude, W. | 25 | 19 | 15 | 20 |
| 12 | Koos, C. | 25 | 12 | 12 | 16 |
| 13 | Leuthold, J. | 25 | 10 | 12 | 11 |
| 14 | Becker, J. | 25 | 9 | 10 | 11 |

**Table 4** Collaboration index result in collaboration age weighted network

| Sl. No. | Name of author | l-index | wl-index | h-degree | c-index |
|---------|----------------|---------|----------|----------|---------|
| 1 | Osterwinter, Heinz | 42 | 40 | 20 | 29 |
| 2 | Hasegawa, T. | 42 | 37 | 27 | 22 |
| 3 | Shimizu, K. | 42 | 30 | 19 | 29 |
| 4 | Yoshino, K. | 42 | 31 | 28 | 22 |
| 5 | Fiore, A. | 36 | 29 | 22 | 24 |
| 6 | Tomkos, I. | 32 | 20 | 22 | 27 |
| 7 | Klonidis, D. | 32 | 21 | 21 | 22 |
| 8 | Kouloumentas, C. | 32 | 27 | 22 | 25 |
| 9 | Lau, Y.Y. | 27 | 10 | 14 | 17 |
| 10 | Gilgenbach, R.M. | 27 | 24 | 20 | 19 |
| 11 | Freude, W. | 25 | 20 | 20 | 20 |
| 12 | Koos, C. | 25 | 14 | 14 | 20 |
| 13 | Leuthold, J. | 25 | 20 | 18 | 21 |
| 14 | Becker, J. | 25 | 21 | 20 | 22 |

In publication count weighted network author **Hasegawa, T., Hasegawa, T. and Tomkos, I.** are key authors in on wl-index, h-degree and c-index, respectively, are shown in Table 3.

In collaboration age weighted network author **Osterwinter, Heinz and Yoshino, K.** are key author in wl-index and h-degree, respectively, and **Osterwinter, Heinz, and Shimizu, K.** are key author in c-index are shown in Table 4.

# 7 Conclusions

In this article, we explored the weighted collaboration network where collaboration weight is like total publication count, total citation count and total collaboration age and evaluate the scientific impact of individuals based on the collaborative index like lobby-index, wl-index, h-degree and c-index. Lobby-index gives equal value in all weighted networks because it does not consider the collaboration weight, but other collaborative index value varies in different weighted network because it depends on the collaboration weight. Here some of the authors having the equal value in all types of weighted networks. The variation in results shows the collaboration impact in scientific evaluation.

# References

1. Wasserman, S., Faust, K.: Social network analysis: Methods and applications, vol. 8. Cambridge University Press (1994)
2. Korn, A., Schubert, A., Telcs, A.: Lobby index in networks. Phys. A: Stat. Mech. Appl. **388** (11), 2221–2226 (2009)
3. Zhao, S.X., Rousseau, R., Fred, Y.Y.: h-degree as a basic measure in weighted networks. J. Inform. **5**(4), 668–677 (2011)
4. Newman, M.E.: Scientifc collaboration networks. I. Network construction and fundamental results. Phys. Rev. E **64**(1), 016131 (2001)
5. Abbasi, A., Altmann, J.: On the correlation between research performance and social network analysis measures applied to research collaboration networks. In: 44th Hawaii International Conference on System Sciences, IEEE, pp. 1–10 (2011)
6. Abbasi, A., Hossain, L., Uddin, S., Rasmussen, K.J.: Evolutionary dynamics of scientific collaboration networks: multi-levels and cross-time analysis. Scientometrics **89**(2), 687–710 (2011)
7. Bihari, A., Pandia, M.K.: Key author analysis in research professionals relationship network using citation indices and centrality. Procedia Comput. Sci. **57**, 606–613 (2015)
8. Arnaboldi, V., Dunbar, R. I., Passarella, A., Conti, M.: Analysis of co-authorship ego networks. In: Advances in Network Science, pp. 82–96. Springer, Berlin (2016)
9. Bihari, A., Tripathi, S., Pandia, M.K.: Key author analysis in research professionals' collaboration network based on MST using centrality measures. In: Proceedings of the Second International Conference on Information and Communication Technology for Competitive Strategies, ACM, p. 118 (2016)
10. Hirsch, J.E.: An index to quantify an individual's scientific research output. Proc. Natl. Acad. Sci. USA **102**(46), 16569–16572 (2005)
11. Yan, X., Zhai, L., Fan, W.: C-index: A weighted network node centrality measure for collaboration competence. J. Informetr. **7**(1), 223–239 (2013)
12. Pandia, M.K., Bihari, A.: Important author analysis in research professional's relationship network based on social network analysis metrics. In: Computational Intelligence in Data Mining, vol. 3, pp. 185–194. Springer, Berlin (2015)
13. Bihari, A., Pandia, M.K.: Eigenvector centrality and its application in research professionals' relationship network. In: Futuristic Trends on Computational Analysis and Knowledge Management, International Conference on, IEEE, pp. 510–514 (2015)

14. Wang, B., Yao, X.: To form a smaller world in the research realm of hierarchical decision models. In: International Conference on Industrial Engineering and Engineering Management, IEEE, pp. 1784–1788 (2011)
15. Wang, B., Yang, J.: To form a smaller world in the research realm of hierarchical decision models. In: Proc. PICMET'11, PICMET (2011)
16. Swart, P.J., Schult, D.A., Hagberg, A.A.: Exploring network structure, dynamics and function using networks. In: Proceedings of the 7th Python in Science Conference (2008)

# Part IX
# Communication Technologies

# Localization in Wireless Sensor Networks Using Invasive Weed Optimization Based on Fuzzy Logic System

**Gaurav Sharma, Ashok Kumar, Prabhleen Singh
and Mohammad Jibran Hafeez**

**Abstract** In this paper, a computationally efficient range-free localization algorithm in 2D plane has been proposed for wireless sensor networks using fuzzy logic system (FLS) and invasive weed optimization (IWO). The received signal strength (RSS) value is used to find out the distance between nodes in the network. To reduce the nonlinearity between RSS values and distances between nodes, and to model the edge weight of the anchor nodes based on the RSS value, FLS is used. Further IWO is used to optimize the edge weights of anchors to achieve the high localization accuracy. Radio propagation is considered as circular for all nodes in the proposed method. It is observed from the simulation results, proposed algorithm performs better in localization accuracy than some existing range free localization algorithms, i.e. centroid method, weighted centroid and range-free localization based on biogeography-based optimization.

**Keywords** Invasive weed optimization · Fuzzy logic system · Localization · Wireless sensor networks

G. Sharma (✉) · A. Kumar · P. Singh · M.J. Hafeez
Department of Electronics and Communication Engineering, National Institute
of Technology Hamirpur, Hamirpur 177005, Himachal Pradesh, India
e-mail: ergaurav209@yahoo.co.in

A. Kumar
e-mail: ashok@nith.ac.in

P. Singh
e-mail: singh.prabhleen123@gmail.com

M.J. Hafeez
e-mail: jibran.hafeez@gmail.com

© Springer Nature Singapore Pte Ltd. 2018
R.K. Choudhary et al. (eds.), *Advanced Computing
and Communication Technologies*, Advances in Intelligent Systems
and Computing 562, https://doi.org/10.1007/978-981-10-4603-2_23

# 1   Introduction

In recent years, due to advancement in microelectronics and wireless communication technologies, most of the industrial interests and research have been attracted towards wireless sensor networks (WSNs). These technologies lead to the development of inexpensive, extremely small, multifunctional and smart sensor nodes [1], which have the capabilities to communicate through a wireless medium.

These nodes comprise a low power processor, sensor(s), small dedicated memory, actuators, battery and a transceiver. Depending on applications, hundreds or thousands of tiny nodes are deployed in the area of interest, which are capable of sensing the environmental parameters such as light, pressure, temperature, humidity, sound, etc. They can exchange data among each other, compute simple tasks on that data and transmit to the central unit called sink node or base station.

Most of the applications such as to monitor the large terrains, to detect forest fire, surveillance of the battlefield, search and rescue operations, target tracking, to observe smart environments, etc. require accurate locations of the nodes. Therefore, finding the location of nodes has become the most important attribute in WSN. Unfortunately, the addition of Global Positioning System (GPS) [2, 3] device is not realistic for a large sensor network [4] because of its production cost factor, extra energy consumption, making the bulky size of nodes, not working in NLOS (non-line of sight) environments, etc [5].

Therefore, alternate of GPS is needed to operate in harsh environments. For the localization process, a small amount of nodes which are aware of their locations known as *anchor nodes or beacon nodes*, are deployed randomly throughout the network with remaining nodes which are unaware of their locations known as *target nodes or location unaware nodes*. The anchor nodes know their locations either by the GPS or by the system administrator or by the manual deployments and normal nodes find their locations with the help of location information of anchors.

Generally, two types of the localization algorithms are there: range-based and range-free. Range-based algorithms show good accuracy at the cost of an extra hardware, that increases the whole cost of the large area network and these algorithms are also affected by the noise and multipath fading [6]. On contrary, range free algorithms are cost effective because connectivity information is only needed between nodes for location estimation. So, most of the researchers pay their attention towards the range free algorithms due to its advantages over range-based algorithms. In this paper, we propose the application of Fuzzy Logic System (FLS) [7] and Invasive Weed Optimization (IWO) [8] on range-free algorithm for 2D randomly distributed nodes in WSNs. Both FLS and IWO performed better and show good results in terms of localization accuracy, the percentage of successfully localized nodes as compared to the existing algorithms.

The paper is structured as follows: Sect. 2 gives the overview of the related techniques of localization. We described our proposed algorithm in Sect. 3. Then, in Sect. 4 simulation parameters and results are discussed, finally, we conclude it in Sect. 5.

## 2 Related Work

In range-free localization, only connectivity information and number of hops between nodes are needed for location calculation. These algorithms are cost-effective than range-based algorithms. But they show poor localization accuracy. Many techniques have been proposed in the literature to improve the localization accuracy of the range-free algorithms. Mainly range-free localization algorithms in literature are centroid [9], approximate-point in triangle (APIT) [10], DV-Hop [11], semi-definite position (SDP) [12], convex position estimation (CPE) [13], etc.

Centroid algorithm is simple and having low computational and communication cost. In this, nodes are placed in grid configuration [9]. It works for those normal nodes, which have at least three neighbour anchors. Normal nodes get their locations by estimating the centroid of the coordinates of the neighbour anchors. Since the accuracy of centroid algorithm depends on the distribution of the anchors and number of anchor nodes in the network. When anchor nodes distribution is regular then it shows good localization accuracy. Also the localization accuracy of centroid is proportional to number of anchor nodes in network.

CPE defines the estimative rectangle (ER), which limits the overlapping region formed by the communication ranges of the neighbour anchor nodes. Then the location of the target node can be determined by calculating the centre of this rectangle. Though CPE is more accurate than Centroid yet its accuracy needs to be improved more because it is not always necessary that the centre of ER is same as the centre of overlapping region [13].

DV-Hop localization algorithm [11] is range-free distributed algorithm. In this algorithm, distance between nodes can be determined using the hop counts between nodes and hop size of the anchor nodes. Since estimated distance between nodes is not so much accurate due to large hop counts and hop size of the anchors. Many improved DV-Hop algorithms have been proposed in the literature but the localization accuracy may vary according to distribution of the anchor nodes.

In [14], an extension of centroid algorithm is presented in terms of adding weights to the coordinates of the anchors to improve the accuracy. These weights are defined according to the distance between anchors and target nodes, i.e. large distance is given a low weight and vice versa. This algorithm is popularly known as weighted centroid localization (WCL) algorithm. Accuracy is improved as compared to centroid algorithm up to some extent.

In [15–17], soft computing approaches and some optimization techniques are applied to range-based and range-free localization algorithm. H-best particle swarm optimization (HPSO) and biogeography-based optimization (BBO) [18] performed better in terms of localization accuracy and computational time. BBO shows the slow convergence with high accuracy and HPSO shows fast convergence with low accuracy. These papers showed the effect of noise variance and anisotropic property of the medium on the localization accuracy.

After considering the shortcomings of the existing algorithms, we propose the application of FLS and IWO in this paper. FLS is used to model the nonlinearity between RSSI and distance between nodes and further IWO is used to minimize the location errors.

## 3  Computational Intelligence (CI) Approaches for Localization in WSN

CI-based algorithms have a bounteous cause of notion for optimization; that is why these algorithms have become prevalent due to their timid computational burden and showing good accuracy. Many optimization techniques have been applied in literature to improve the localization accuracy, i.e. Genetic Algorithm (GA), PSO, HPSO, BBO, Firefly Algorithm (FA), Differential Evolution (DE), etc. Some of these optimization techniques show fast convergence but low accuracy and vice versa. For this purpose, we propose IWO which shows fast convergence and high accuracy. The following sections present an overview of fuzzy logic system (FLS) and IWO.

### 3.1  Fuzzy Logic System

The proposal of fuzzy logic was firstly given by Zadeh [7] in the year of 1965. Fuzzy logic is a form of fuzzy set theory which relates to classes of objects with unsharp boundaries in which membership is a matter of degree. Precise inputs are not required in FLS, but the system complexity increases with number of inputs and outputs.

The fuzzy inference system (FIS) is a fuzzy logic system in which inputs are processed according to a list of if-then statements called rules. A FIS consists of a fuzzifier unit some fuzzy IF-THEN rules knowledge base unit, an inference decision-making unit and a defuzzifier unit as shown in Fig. 1.

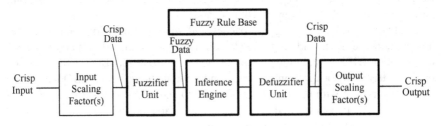

**Fig. 1** Fuzzy logic system (FLS)

The fuzzy model composed of following structure of the rules:

$$R_k: \text{IF } x_1 \text{ is } A_1 \text{ and } A_2 \text{ THEN } y \text{ is } W_k$$

where $R_k$ is the $k$th rule, $x_1$ and $x_2$ are the input variables, $y$ is the output variable, $A_1$ and $A_2$ denote membership functions for inputs and similarly $W_k$ is for output of Mamdani inference system. We use Mamdani inference system in this paper due to its solid defuzzification that gives the results FLS in the consistent form.

## 3.2 Invasive Weed Optimization (IWO)

This stochastic optimization algorithm was firstly introduced by Mehrabian and Lucus in 2006 [8] after inspiring from formation of colonies due to weeds. It is shown that this optimization technique not only outperforms than other optimizers like PSO, GA, BBO, etc. but is also capable of handling some new localization–optimization problems in WSN. There are some following basic properties is considered for the process of the colonizing behaviour of the weeds in IWO [8].

1. Initially, the seeds are being spread over the entire search space.
2. Each seed in the search area produces a complete plant and depending on the fitness each plant yields seeds.
3. The produced seeds are again being distributed over the entire search space and grow to new plants again.
4. When the number of plants reaches to maximum value, then the process ends. The survival of plants depends on their fitness. According to this process, hopefully the fittest plant is near to the optimal solution.

The detailed process of IWO is divided in four parts:

(i)   Initialize a population
(ii)  Reproduction
(iii) Spatial dispersal
(iv)  Competitive exclusion.

For detailed process of IWO, readers are advised to go through [8].

## 4  FLS and IWO-Based Node Localization

The received signal strength (RSS) value is used to find out the distance between nodes in the network. To reduce the nonlinearity between RSS values and distances between nodes, and to model the edge weight of the anchor nodes based on the RSS value, FLS is used. In this section, we present the proposed approach for range free node localization in WSN for 2D scenario. FLS is used to design the edge weights and IWO is applied to optimize these edge weights to achieve the better localization accuracy. Sensor nodes are randomly distributed over the 2D plane. Some assumptions are made for the proposed algorithm, as follows:

1. As discussed in Sect. 1, that location of anchor nodes' is known either by manual deployment or by GPS. Anchor nodes send a beacon message throughout the network which contains its location and RSS (received signal strength) value. Each target node listens the beacon signal from their adjacent anchor nodes and collects RSS value and their locations.
2. Radio propagation is considered as circular and transmission ranges of all nodes are considered same. In practice, it is not purely circular in real scenarios.

For 2D localization in WSN, following steps are performed:

1. Each target node maintained a list of RSS values from their adjacent anchor nodes after receiving the beacon signal.
2. Check if the number of neighbour anchor nodes to a particular target node $\geq 3$ or not, if $\geq 3$, then the target node is considered as localizable node.
3. Calculate the edge weights of anchor nodes to their RSS value. FLS is used to model these edge weights.
4. After calculating the edge weights, calculate positions of the target nodes according to the formula given in [14] as follows in Eq. (1):

$$(x_t, y_t) = \left[ \frac{(w_1 x_1) + \cdots (w_n x_n)}{\sum_{i=1}^{k} w_i}, \frac{(w_1 y_1) + \cdots (w_n y_n)}{\sum_{i=1}^{k} w_i} \right] \tag{1}$$

where $(x_t, y_t)$ is the coordinate of $t$th target node, $(x_1, y_1)$, $(x_2, y_2)$ ... $(x_n, y_n)$ are the coordinates of anchor nodes and $w_1, w_2 \dots w_n$ are the weights between anchor and target nodes.

In this paper, Mamdani fuzzy model for fuzzy rule base is used and further weights are optimized by IWO.

## 4.1 Fuzzy Modelling with Edge Weight Optimization Using IWO

When anchor nodes transmit beacon signal throughout the network, each target node collects the beacon signal containing RSS value and location of the anchor. This RSS value varies according to environmental factors. It is not fixed for a particular distance, i.e. it is uncertain but it gives the clue to find out the distance. FLS is used to overcome this uncertainty and nonlinearity between RSS and distance.

Input variable in rule base of Mamdani fuzzy model is RSS value of the anchor and it is taken in the interval of [0, $RSS_{max}$], where $RSS_{max}$ is 100 dB (i.e. maximum RSS value). Five membership functions are used to map input variable RSS, i.e. VLOW, LOW, MEDIUM, HIGH, VHIGH as shown in Fig. 2. Edge weight is

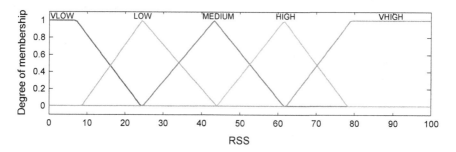

**Fig. 2** Fuzzy membership functions of RSSI

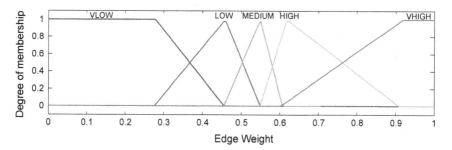

**Fig. 3** Fuzzy membership functions of edge weights generated by IWO

**Table 1** Rule base of edge weight

| S. No. | Antecedent | Consequent |
|---|---|---|
| 1 | IF RSS is VLOW | THEN edge weight is VLOW |
| 2 | IF RSS is LOW | THEN edge weight is LOW |
| 3 | IF RSS is MEDIUM | THEN edge weight is MEDIUM |
| 4 | IF RSS is HIGH | THEN edge weight is HIGH |
| 5 | IF RSS is VHIGH | THEN edge weight is VHIGH |

taken as output variable and the values used for edge weight is in the interval of $[0, w_{max}]$, where $w_{max}$ is 1 (i.e. maximum edge weight). Further edge weights are optimized using IWO. Like RSS, edge weights are also mapped to five membership functions as shown in Fig. 3. The rule base of five rules is shown in Table 1.

## 5 Simulation Results and Discussion

To evaluate the performance of proposed method, we simulate our intelligent localization algorithm in MATLAB environment with total of 100 sensor nodes, which are randomly distributed in a field of $100 \times 100$ m$^2$ as shown in Fig. 4. Communication range of all the nodes is kept same as 35 m. If the distance between

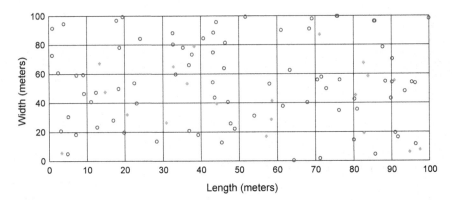

**Fig. 4** An example of nodes distribution

the nodes is less than communication range, then they can communicate with each other. We are assumed the radio propagation is perfectly circular, i.e. DOI (degree of irregularity) value is 0, it means the effect of noise and fading is not considered. Twenty independent trials of IWO are conducted to achieve the optimal values of edge weights as shown in Fig. 3.

To evaluate the results of our proposed algorithm (Range free invasive weed optimization using fuzzy logic (RFIWO + Fuzzy)), we run it in the MATLAB and compare it with the existing algorithms of centroid [9], weighted centroid [14] and range free biogeography-based optimization using fuzzy logic (RFBBO + Fuzzy) [17].

## 5.1 IWO Parameters

IWO parameters to achieve the optimal values of edge weights in our proposed work are shown in Table 2.

Localization error (LE) and average location error (ALE) is calculated using Eqs. (2) and (3), respectively, according to the distance between actual coordinates of target nodes and estimated coordinates of target nodes.

**Table 2** IWO parameters

| S. No. | Parameters | Optimal value |
|--------|-----------|---------------|
| 1 | Maximum population size | 150 |
| 2 | Maximum number of seeds | 5 |
| 3 | Minimum number of seeds | 0 |
| 4 | Non-linear modulation index ($n$) | 3 |
| 5 | DOI | 0 |
| 6 | Max. no. of iterations | 100 |
| 7 | No. of trials | 20 |
| 8 | $\sigma_{final}$ | 10% of search range |
| 9 | $\sigma_{initial}$ | 0.004% of search range |

$$LE = \sqrt{(x_e - x_a)^2 + (y_e - y_a)^2} \qquad (2)$$

$$ALE = \frac{\sqrt{(x_e - x_a)^2 + (y_e - y_a)^2}}{n} \qquad (3)$$

where $(x_e, y_e)$ are the estimated coordinates of target nodes and $(x_a, y_a)$ are the corresponding actual coordinates of target nodes and $n$ is the number of target nodes. Maximum, minimum, and average localization errors of all mentioned algorithms are presented in Table 3.

Figure 5 shows the localization errors of the centroid, weighted centroid, RFBBO + Fuzzy and our proposed algorithm (RFIWO + Fuzzy). From these

**Table 3** Localization errors of different localization algorithm

| Range free methods | Max. location error | Min. location error | Avg. location error |
|---|---|---|---|
| Centroid localization | 3.284 | 0.0955 | 1.785 |
| Weighted centroid localization | 2.247 | 0.0580 | 1.152 |
| RFBBO + Fuzzy | 0.431 | 0.0003 | 0.258 |
| RFIWO + Fuzzy | 0.384 | 0.0001 | 0.192 |

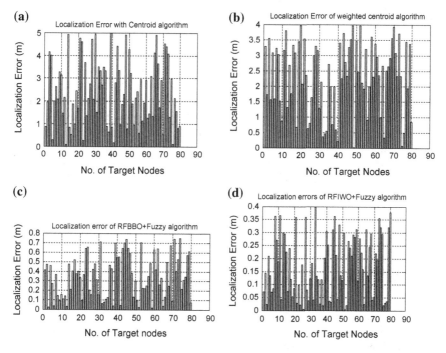

**Fig. 5** Localization error of each target node **a** centroid algorithm, **b** weighted centroid algorithm, **c** RFBBO + Fuzzy algorithm, **d** RFIWO + Fuzzy algorithm (proposed)

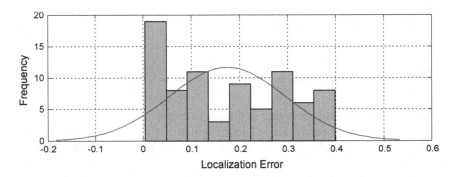

**Fig. 6** Frequency of error occurrence of RFIWO + Fuzzy algorithm

simulation results, we analyzed that our proposed algorithm outperforms the other existing algorithms in terms of localization accuracy. Figure 6 shows the frequency of errors occurrence in proposed method. We can see from this result that means localization error is near by 0.192 m. In this paper, it is observed that the proposed algorithm performs better as compared to other existing range free localization algorithms.

## 6 Conclusion and Future Scope

To improve the localization accuracy in WSN, a distributed range-free localization algorithm for 2D plane using fuzzy logic system (FLS) and invasive weed optimization (IWO) referred as RFIWO + Fuzzy is proposed in this paper. Proposed algorithm does not need any extra hardware to get the distance information. Only RSS value is sufficient for calculating the distance between nodes. This RSS value is used to define the edge weights of the anchor nodes in FLS. Further to minimize the localization errors, IWO is used to optimize the edge weights of anchor nodes. Simulation results show that the proposed algorithm performs better as compared to some existing range-free localization algorithms, i.e. centroid, weighted centroid, range free BBO using fuzzy logic. The proposed method can be extended for 3D space with consideration of anisotropic property of the propagation.

## References

1. Akyildiz, I.F., Su, W., Sankarasubramaniam, Y., Cayirci, E.E.: A survey on sensor network. IEEE Commun. Mag. **40**(8), 102–105 (2002)
2. Hofmann-Wellenfof, B., Lichtenegger, H., Collins, J.: Global Positioning System: Theory and Practice. Springer, Berlin (1993)

3. Djuknic, G.M., Richton, R.E.: Geolocation and assisted GPS. Computer **34**(2), 123–125 (2001)
4. Niculescu, D., Nath, B.: Ad Hoc positioning system (APS). In: Proceedings Global Telecommunication Conference (Globecom'01), vol. 1, pp. 292–293 (2001)
5. Boukerchie, A., Oliveria, H.A.B.F., Nakamura, E.F., Loureiro, A.A.F.: Localization systems for wireless sensor networks. IEEE Wirel. Commun. **14**(6), 6–12 (2007)
6. Ward, A., Jones, A., Hopper, A.: A new location technique for the active office. IEEE Pers. Commun. **4**(5), 42–47 (1997)
7. Zadeh, L.: Fuzzy logic: computing with words. IEEE Trans. Fuzzy Syst. **4**(2), 103–111 (1996)
8. Mehrabian, A.R., Lucas, C.: A novel numerical optimization algorithm inspired from weed colonization. Ecol. Inform. **1**, 355–366 (2006)
9. Bulusu, N., Heidemann, J., Estrin, D.: GPS-less low-cost outdoor localization for very small devices. IEEE Pers. Commun. **7**(5), 28–34 (2000)
10. Zhou, Y., Ao, X., Xia, S.: An improved APIT node self-localization algorithm in WSN. In: 7th World Congress on Intelligent Control and Automation (WCICA 2008), Chongqing, China, pp. 7582–7586 (2008)
11. Gao, G., Lei, L.: An improved node localization algorithm based on DV-HOP in WSN. In: 2nd International Conference on Advanced Computer Control (ICACC), IEEE, Shenyang, China, vol. 4, pp. 321–324 (2010)
12. Bachrach, J., Taylor, C.: Localization in sensor networks. In: Handbook of Sensor Networks, pp. 277–310 (2005)
13. Doherty, L., Kristofer, S.J., Laurent, E.: Convex position estimation in wireless sensor networks. In: Proceeding of INFOCOM, Anchorage, AK, vol. 3, pp. 1655–1663 (2001)
14. Kim, S.Y., Kwon, O.H.: Location estimation-based on edge weights in wireless sensor networks. Inform. Commun. Soc. **30**(10A) (2005)
15. Kumar, A., Khosla, A., Saini, J., Singh, S.: Meta-heuristic range-based node localization algorithm for wireless sensor networks. In: International Conference on Localization and GNSS (ICL-GNSS), IEEE, Munich, Germany, pp. 1–7 (2012)
16. Kumar, A., Khosla, A., Saini, J., Singh, S.: Computational intelligence-based algorithm for node localization in wireless sensor networks. In: 6th IEEE International Conference on Intelligent Systems (IS), IEEE, Bulgario, pp. 431–438 (2012)
17. Kumar, A., Khosla, A., Saini, J., Singh, S.: Range-free 3D node localization in anisotropic wireless sensor networks. Appl. Soft Comput. **34**, 438–448 (2015)
18. Simon, D.: Biogeography-based optimization. IEEE Trans. Evol. Comput. **12**(6), 702–713 (2008)

# Genetic Algorithm-Based Heuristic for Solving Target Coverage Problem in Wireless Sensor Networks

**Manju, Deepti Singh, Satish Chand and Bijendra Kumar**

**Abstract** Target coverage problem in the wireless sensor networks schedule sensors into subsets such that all the targets are monitored by each subset. Existing heuristics to solve this problem aims to maximize the total network lifetime. In last few decades, genetic algorithm-based methods have been proved more suitable to solve such optimization problems. In this paper, we propose a solution heuristic for target coverage problem which is based on genetic algorithm approach. The simulation results show that the proposed heuristic outperforms the existing methods.

**Keywords** Target coverage · Genetic algorithm · Meta-heuristic · Network lifetime

## 1 Introduction

In Wireless Sensor Networks (WSNs), tiny devices called sensors are densely deployed to provide continuous surveillance for many applications [1]. These sensor nodes are provided with small battery device which is not rechargeable and sometimes not even replaceable. In order to maximize their battery usages, we need to optimize the sensors usages in energy-efficient manner. As discussed in literature

Manju (✉) · D. Singh · B. Kumar
Netaji Subhash Institute of Technology, Sector-3, Dwarka,
New Delhi 110078, India
e-mail: manju.nunia@gmail.com

D. Singh
e-mail: deepti19singh@gmail.com

B. Kumar
e-mail: bizender@gmail.com

S. Chand
School of Computer and System Sciences, Jawaharlal Nehru University,
New Delhi 110067, India
e-mail: schand20@gmail.com

© Springer Nature Singapore Pte Ltd. 2018
R.K. Choudhary et al. (eds.), *Advanced Computing
and Communication Technologies*, Advances in Intelligent Systems
and Computing 562, https://doi.org/10.1007/978-981-10-4603-2_24

[2–6], the target coverage problem is to maximize the global network lifetime while covering all the targets for maximum period. The very blind approach to do this is to activate all the sensors deployed in the sensor field to ensure full coverage. But, in this way, the network will not be functional for longer period which results in decreased network lifetime. Thus, lots of work [2–9] has been done where only a subset of sensors is activated at a time which ensures full targets coverage. Theses subsets are known as sensor covers. Each sensor cover is activated for a fixed duration (called working time). The network lifetime is termed as the sum of all the sensor covers activation times. Hence, the objective behind all the works [2–9] so far is to find maximum possible such sensor covers to maximize the network lifetime. In the next section, we discuss the existing research works done to solve the target coverage problem and identify the gap between these works to design a new solution heuristic.

## 2  Existing Work

To solve the target coverage problem, Slijepcevic and Potkonjak [2] discussed a greedy heuristic which generates non-disjoint sensor covers where s sensor can be part of only single sensor cover. Due to this restriction on sensors' participation, the network lifetime obtained by this heuristic is not optimal. To improve the network lifetime, Cardei et al. [3] proposed an approximation algorithm where generated sensor covers are non-disjoint in which a sensor can take part in more than one sensor cover. Carde et al. [3] proved that the target coverage problem belongs to NP-Complete class. Further, works in [4–7] also presented heuristic solution to maximize the total network lifetime. Since last few decades, several studies have shown [8–11] that genetic algorithms-based heuristics are more promising over the conventional methods [2–7] to solve the target coverage problem. This is so because nonlinear optimization problem are handled well by genetic algorithm-based approaches [8]. Genetic algorithm starts randomly generated initial population where each solution is represented using binary string of genes called chromosomes. There are fixed number of elements (called genes) in each chromosome. A fitness function is used to determine the quality of a chromosome. In case of target coverage problem, the number of elements in a chromosome is equals to the number of sensors in the given network. Here, a chromosome represents a sensor cover. Gupta et al. [8] addressed a GA-based heuristic to solve target coverage problem. To maximize the network lifetime, this heuristic try to minimize the number of positions selected to place the sensor in the given networks. The heuristic also fulfil the $k$-coverage and $m$-connectivity requirements where each target is said to be covered if it covered by at least $k$-number of sensors and all the sensors are connected to $m$-different sensors in the network. Carrabs et al. [9] also addressed a genetic algorithm-based metaheuristic which provides full as well as partial coverage while solving target coverage problem. In this heuristic, the fitness function tries to reduce the cardinality of the generated sensor cover to maximize the gain in network lifetime. Further, Rebai et al. [10] addressed another genetic algorithm which finds minimum locations to place sensor. The generated sensor cover by this method also ensures connectivity within the network.

In this paper, we propose a novel genetic algorithm-based metaheuristic which solve the target coverage problem and maximize the network lifetime.

## 3 Proposed Heuristic

Our proposed heuristic is based on genetic algorithm approach, thus, it consists of all the steps involved in genetic algorithm method. The major steps include initial population, chromosome representation, fitness function, crossover and the mutation operations. Here we discuss how these major operations are performed in our proposed heuristic to solve the *target* coverage problem.

### 3.1 Chromosome Representation

In the proposed heuristic, the chromosome is a string of zeros and ones of length equals to number of sensors. In this string, the value of $i$th gene is 1 when the $i$th sensor is active in the current sensor cover otherwise its 0. Now we illustrate the same with the example as shown in Fig. 1. Here, we take a sensor network which consists of with 4 targets $T = \{t_1, t_2,..., t_4\}$ and 6 sensors $S = \{s_1, s_2,..., s_6\}$. Therefore, the length of the chromosome is 6 for the given network. The gene values are 1 and 0 as shown in Fig. 1. At position 1, the gene value is 1 which shows that sensor $s_1$ is active in the sensor cover. Similarly, the sensor $s_3$ is not active as the gene value at position 3 is 0. Thus, in the given sensor cover, $s_1$, $s_3$ and $s_5$ are active sensors whereas $s_2$ and $s_4$ are in sleep mode.

#### 3.1.1 Initial Population Generation

In the proposed heuristic, initial population is the set of chromosomes which are chosen by the selection process at the start of process. In order to do that, the chromosomes are randomly generated in a fixed amount. As chromosomes are the binary string of 0 and 1, we propose the following method (*Algorithm 1*) to generate them as given below.

```
Algorithm 1: Initial Population (U) Generation
Input: number of sensors (S), size of initial population (P)
Output: initial population consists of P chromosomes
1. Boolean STRING[P];
2. initialize U = {};
3. for i =1 to P
4.      for j=1 to S
5.           STRING [j] = random( ) %2
6.      end for
7.    U=U ∪ STRING
8. end for
```

**Fig. 1** Chromosome representation

| Gene | 1 | 2 | 3 | 4 | 5 | 6 |
|------|---|---|---|---|---|---|
| value → | 1 | 0 | 1 | 0 | 1 | 0 |

Sensors →

### 3.1.2 Fitness Function

The fitness function is used to determine the quality of newly generated chromosome. A fitness function consists of certain objectives which should be fulfilled every time. As we know that the targets which are covered by least number of sensors decided the upper bound on the network lifetime. Such targets are known as critical targets. In order to maximize the global network lifetime, our proposed metaheuristic select only least required sensors with highest remaining energy in each sensor cover.

*Objective: Select least number of sensors*: To generate sensor cover, we should select minimum number of sensors say, $K$, therefore, we need to minimize the select sensor in each cover set. Thus, our objective is as given below:

$$\text{Minimize} \frac{K}{S} \tag{1}$$

As sensors are selected with highest remaining energy, therefore, once the sensor cover is generated, we need to minimize the sensor cover by removing redundant sensors. Next we discuss the selection operation followed by the proposed heuristic.

### 3.1.3 Selection

Our heuristic selects only valid chromosomes using various Roulette-wheel selections. In this method, the chromosomes with highest fitness value are selected. Then, selected chromosomes follow crossover operation to produce new chromosomes (known as child chromosome).

### 3.1.4   Crossover

In genetic algorithms, the crossover operation is aimed to create new chromosomes using previously existing chromosomes from the initial population. For this reproduction process, two chromosomes (i.e. parent chromosome) are selected through breeding selection process to produce new chromosomes (called child chromosomes). The child chromosome is inherited with good properties from the parent chromosomes. In this process, if newly generated child chromosome is not different from the existing chromosomes in the initial population pool, then, the child chromosome is either discarded or goes through two-point crossover.

And, to produce more new child chromosomes, the crossover point is shifted by one bit ahead as compared to the earlier crossover point. There are many types of crossover operations that exist like one-point crossover, uniform crossover and two-point crossover. The easiest crossover operator is single point crossover where a single point is randomly selected to divide the role of parent chromosomes. Here Fig. 2a shows the good example of mating by parent chromosomes to produce the child chromosome using single point crossover. In this figure, there are two parent chromosomes namely P1 (1 0 1 0) and P2 (1 1 0 1). The single crossover point is set after second gene value. Therefore, gene values after this point will be exchanged for both the parents to produce child chromosome namely child 1 (1 0 0 1) and child 2 (1 1 1 0). As shown here, both the child chromosomes are different from the parent chromosomes P1 and P2.

Now, we see the scenario where produced child chromosome is same as one of the parent chromosomes as shown in Fig. 2b. In this case, the produced child chromosomes (child 1 and child 2) are same as parent chromosomes P1 (1 0 1 0 1) and P2 (1 1 1 0 1). If this is the case, then both the child chromosomes are discarded and new chromosomes are generated using different crossover points.

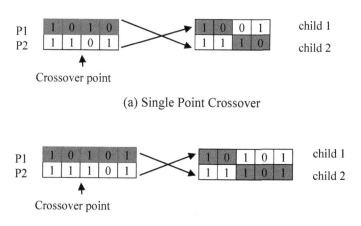

(a) Single Point Crossover

(b) Single Point Crossover Operation

**Fig. 2**  **a** Single point crossover, **b** single point crossover operation

**Mutation**

While producing child chromosomes using crossover operation, there may be the case that produced child chromosomes may not be valid. A valid child chromosome is that which cover all the targets (i.e. it is a sensor cover). To make an invalid chromosome valid, the mutation operation is applied. In this operation, a gene value is picked randomly and changed to 1 from 0. Thus it is the complete set of operations which are followed by the proposed heuristic to solve the target coverage problem.

In the next section, we perform a wide set of simulations to claim the superiority of the proposed genetic algorithm-based heuristic with the non-genetic methods [2, 3, 6].

# 4 Simulation

In this section, we perform a detailed set of simulations to compare the performance of the proposed metaheuristic with some other existing methods by Slijepcevic and Potkonjak [2], Cardei et al. [3] and Bajaj and Manju [6]. In order to compare these methods, we consider prime parameters like network lifetime verses number of sensor nodes, like network lifetime versus number of targets, sensing ranges. All the simulation test scenarios are randomly generated and each value is an average of 30 instances in total. We take 50 chromosomes in initial population. We follow the radio model in [3] to compute the energy consumption during transmission and receiving the data by base station. The total energy consumption while sending and receiving an **n**-bit data packet at distance **d** is given in Eqs. (2) and (3) as follows:

$$E_{tx} = P \times \left( E_{amp} + \varepsilon_{fs} \times d^n \right) \qquad (2)$$

$$E_{rx} = P \times E_{amp} \qquad (3)$$

Now we see the performance of various methods in [2, 3, 6] and the proposed heuristic. First we take 50 targets and take sensors between 20 and 150. The network is assumed to be homogenous where each sensors has sensing range (70 m) and energy level ($E_i = 1$ J). The simulation results for this scenario are shown in Fig. 3. It is clearly evident from Fig. 3 that network lifetime increases while increasing the sensor nodes in the same area. This happens so because with more sensors network will be denser which results in prolonged coverage for the same targets. It is also evident from Fig. 3 that the proposed metaheuristic outperforms methods in [2, 3, 6]. Thus, we can say that genetic algorithm-based techniques are more promising over conventional schemes.

Next we performed another experiment in the same network to see how network lifetime effects while changing targets in the fixed area. To do this, we take fixed number of sensors (150) and varying targets from 20 to 100. Figure 4 shows the outcomes of this experiment.

**Fig. 3** Network lifetime versus number of sensors

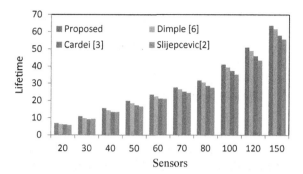

**Fig. 4** Network lifetime versus number of targets

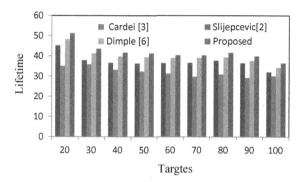

**Fig. 5** Network lifetime versus sensing range

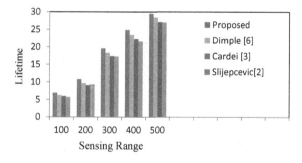

As shown in Fig. 4, there is sharp decrement in network lifetime obtained. This is due to the fact that when number of targets increases in the same area then, more sensor nodes is required to cover all the targets which in turn results in lesser network lifetime. It is also evident from Fig. 4 that the proposed genetic algorithm-based metaheuristic outperforms the conventional methods [2, 3, 6].

Next, we see how the sensing range affects the network lifetime. To perform this experiment, we take fix targets (50) and sensors (200). We vary the sensing range between 100 and 500 m. Rest of the network parameters is same as previous experiments. The simulation outcome for this experiment is shown in Fig. 5. It is observed in Fig. 5 that network lifetime increases while sensing range increases.

It happens so because with longer sensing range, targets can be covered by lesser number of sensors which in turn increases the network lifetime. Again, the proposed metaheuristic outperforms the existing methods [2, 3, 6].

## 5    Conclusion

In this paper, a Genetic Algorithm-based metaheuristic, to maximize the global network lifetime for the target coverage problem in the given networks has been proposed We aim to obtain the network lifetime near to the optimal upper bound by selecting only minimum number of sensors with highest residual energy. Our proposed heuristic outperforms some existing approaches [2, 3, 6]. The simulations have been carried out with various parameters to claim the superiority. In future, this work might be extended to the next level including some more variants like Q-Coverage, Variable radii Coverage and Connected Coverage.

## References

1. Akyildiz, I.F., Su, W., Sankarasubramaniam, Y., Cayirci, E.: A survey on sensor networks. IEEE Commun. Mag. **40**(8), 102–114 (2002)
2. Slijepcevic, S., Potkonjak, M.: Power efficient organization of wireless sensor networks. In: Proceedings of the IEEE International Conference on Communications, 472–476 (2001)
3. Cardei, M., Thai, M.T., Li, Y., Wu, W.: Energy-efficient target coverage in wireless sensor networks. In: Proceedings of the 24th Conference of the IEEE Communications Society, vol. 3, 1976–1984 (2005)
4. Manju, N., Chand, S., Kumar, B.: Maximizing network lifetime for target coverage problem in wireless sensor networks, IET Wirel. Sens. Syst. doi:10.1049/iet-wss0094 (2015)
5. Pujari, A.K., Mini, S., Padhi, T., Sahoo, P.: Polyhedral approach for lifetime maximization of target coverage problem. In: Proceedings of the 2015 International Conference on Distributed Computing and Networking, pp. 1–8 (2015)
6. Dimple, M.: Maximum Coverage Heuristics (MCH) for target coverage problem in wireless sensor networks. In: 4th IEEE International Advance Computing Conference (IACC2014), pp. 300–305 (2014)
7. Chaudhary, M., Pujari, A.K.: Q-coverage problem in wireless sensor networks. In: Proceedings of the 2009 International Conference on Distributed Computing and Networking, pp. 325–330 (2009)
8. Gupta, S.K., Kuilab, P., Jana, P.K.: Genetic algorithm approach for k-coverage and m-connected node placement in target based wireless sensor networks. Comput. Electr. Eng. 1–13 (2015)
9. Cerulli, F.C.R., D'Ambrosio, C., Raiconi, A.: A hybrid exact approach for maximizing lifetime in sensor networks with complete and partial coverage constraints. J. Netw. Comput. Appl. **58**, 12–22 (2015)
10. Rebai, M., Leberre, M., Snoussi, H., Hnaien, F., Khoukhi, L.: Sensor deployment optimization methods to achieve both coverage and connectivity in wireless sensor networks. Comput. Oper. Res. **59**, 11–21 (2015)
11. Raiconi, A., Gentili, M.: Exact and metaheuristic approaches to extend lifetime and maintain connectivity in wireless sensors networks. Lecture Notes in Computer Science, vol. 6701, 607–619. Springer, Berlin/Heidelberg (2011)

# Target Coverage Heuristics in Wireless Sensor Networks

Manju, Deepti Singh, Satish Chand and Bijendra Kumar

**Abstract** In order to cover a fixed set of targets within a wireless sensor networks, sensor nodes are closely deployed. Since, sensors are heavily deployed, instead of all sensors, few sensors are sufficient to monitor full target set. Therefore, to maximize the total network lifetime for covering all the targets (target coverage problem), sensor nodes are partitioned into cover sets. Each cover set is sufficiently enough to monitor the complete target set. Then the cover sets are activated sequentially for a fixed duration. The summation of all the covers duration is called network lifetime. More number of such cover set results in longer global network lifetime. To do this, in this work, we propose an energy optimal heuristic which generates maximum possible cover sets to reach the upper bound calculated for network lifetime.

**Keywords** Target coverage · Network lifetime · Upper bound · Cover set

## 1 Introduction

Wireless sensor network mainly consists of tiny, low-powered sensors incorporated with a small microprocessor, transceiver, and memory resource to make it functional. Mainly, sensor networks are used in area like wild life monitoring, fire detection, land

Manju (✉) · D. Singh · B. Kumar
Netaji Subhash Institute of Technology, Sector-3, Dwarka,
New Delhi 110078, India
e-mail: manju.nunia@gmail.com

D. Singh
e-mail: deepti19singh@gmail.com

B. Kumar
e-mail: bizender@gmail.com

S. Chand
School of Computer and System Sciences, Jawaharlal Nehru University,
New Delhi 110067, India
e-mail: schand20@gmail.com

© Springer Nature Singapore Pte Ltd. 2018
R.K. Choudhary et al. (eds.), *Advanced Computing
and Communication Technologies*, Advances in Intelligent Systems
and Computing 562, https://doi.org/10.1007/978-981-10-4603-2_25

265

sliding detection, and military application [1–3] where human accessibility is not possible. To facilitate these applications, sensors are densely deployed to provide continuous surveillance. Sensor nodes are equipped with small battery device, which is not rechargeable. Thus, we need to optimize this limited source. In this paper, our main focus is on sensor scheduling for target coverage problem. Here, we assume that sensor network is densely deployed. One way to increase the total network lifetime is to put redundant sensors to *Sleep* mode. In order to follow this strategy, activate only a subset of sensor (say cover set) to monitor the complete set of targets. Then activate these cover set consecutively in the given environment which in turn maximizes the network lifetime. Many works addressed this coverage problem (target coverage) as maximum network lifetime problem (MLP) [4, 5]. The main objective of the MLP is to maximize the functional duration of a deployed network by guaranteeing the coverage of fix number of targets. Recent works shows bit more interest to use the exact approaches while solving optimization problems in wireless sensor networks [6–10]. The column generation approach has been proved to be better for solving the maximum network lifetime problem. The column generation theory basically divides the original problem into two parts called master and a subproblem. Here, the master problem is a class of linear programming whose solution is denoting the lower bound for network lifetime achievable. In order to find upper bound for network lifetime, the subproblem is used. The subproblem is the integer programming (IP) problem, thus, belongs to the NP-complete class which in turn not solvable in polynomial time. Thus to solve this subproblem, either exact [5–7] or heuristic approach can be used. Mostly, heuristic approaches are discussed by many of the researchers in [2–4, 8–10].

## 2 Related Work

In WSNs, coverage problems are broadly classified as area coverage and target coverage [4]. The objective of the area coverage problem is to cover certain area, while in target coverage [5] a set of target should be covered inside a deployed network. There have been lots of works done on both the coverage problems [2–11]. Here we discuss target coverage problem. In this problem, targets are deployed within fixed sensing area, and the objective is to monitor these targets for the longest possible duration. In case of disjoint cover set, sensors are restricted to participate in single cover set [4–6]; therefore, the total lifetime of the network is calculated by multiplying the calculated number of cover sets with a fixed amount (working time of cover set). Non-disjoint cover sets [8, 9] allows each sensor to participate in multiple cover set and make sure that none of the sensor is used more than its initial energy assigned with. The works done in [5–7, 12] solves the target coverage using exact methods. Slijepcevic and Potkonjak [5] first convert the area coverage into equivalent target coverage problem and presented this as maximum network lifetime (MPL) problem. They solved this MLP using exact approach based on column generation strategy. Gu et al. [6] uses column generation (CG) technique to convert the MLP into pattern-based representation where a cover set (pattern) is selected to cover the complete targets set. This approach finds near optimal solution which can further improved by applying heuristic methods embedded in column generation. Work in [7] also solves this MLP by generating its columns (sensor covers) as when necessary by using exact approach based on CG. The exact approach [7] is suitable for medium size instances because the columns of MLP increase exponentially with sensors. As discussed earlier in Sect. 1, solving the coverage MLP using exact approach based on CG techniques is constrained by many parameters like number of sensors, optimal upper bound, and increasing number of columns (covers) of MLP. Thus, it is advisable to solve it heuristic-based column generation technique. All the works in [2–4, 8–11] presented many heuristic approaches for the target coverage problem which achieves the near optimal upper bound on network lifetime. Due to the presence of faulty nodes in the deployed sensor network, another variant of target coverage is introduced called Q-coverage [9, 10]. In Q-coverage, every single target must be covered by at least Q-number of sensors in each cover set to support QoS parameters. Therefore, faulty nodes could not affect the functionality of given network as targets are continuously covered with more than one sensor. In [9], Pujari et al. presented a heuristic which generate Q-covers where Q-cover is capable of monitoring each target by Q-number of sensors. In order to select sensors to be part of Q-cover, energy is considered as prime metric. Therefore, high remaining energy sensors are selected first. Once Q-cover is generated, redundant sensors are removed to prolong network lifetime. Mini et al. [10] presented another heuristic which generate cover sets to support Q-coverage by considering energy and coverage as prime parameters.

## 3 Proposed Heuristic

Here we first define the target coverage problem with linear programming formulation, and then we propose the new minimal-heuristic for the same problem.

**Target Coverage Problem**: We consider a fixed sensing area, where a set $T = \{t_1, t_2,...t_n\}$ of $n$ number of target randomly placed. In order to monitor these targets, $m$ number of sensor $S = \{s_1, s_2,...s_m\}$ with fix sensing range is deployed. Each sensor $s$ is assigned an initial battery E. A target $t_r$, $1 \leq r \leq n$, is covered if and only if it comes under the sensing range of a sensor $s_i$, $1 \leq i \leq m$. A sensor can alternate between active or sleep state. We try to find out maximum possible either disjoint **or** non-disjoint cover sets. Let $P$ be the set of the entire cover sets, i.e., $P = \{P_1, P_2,..., P_{\max}\}$. The working time for cover set $P$, $X(P)$, is defined as the energy of the sensor in $K$ which has the minimum battery, i.e., $X(P) = \text{Min}_{s_i \in P}(E_i)$. The Linear Programming (LP) formulation of the coverage problem in given WSN is as follows:

$$
\begin{aligned}
\text{Maximize} \quad & \sum_p x_p \\
\text{subject to} \quad & \sum_p M_{ip} x_p \leq E_i \quad \text{for all sensors} \quad s_i \\
& x_p \geq 0, \quad \text{for all cover-set } C_p.
\end{aligned}
\tag{1}
$$

where $M$ is the constraint matrix defined as follows:

$$M_{ij} = \begin{cases} 1, & \text{if sensor } s_i \text{ is in covrset } K_j \\ 0, & \text{otherwise} \end{cases} \tag{2}$$

As discussed earlier, deployed network is dense, which results in more number of columns of matrix $M$ (cover sets). Therefore, none of the algorithm for solving the given linear programming can be applied as $M$ is not known explicitly. Another way to solve this problem is to generate cover sets (columns of $M$) when required by making use of column generation method applied for linear programming. Authors in [4] proposed the greedy based heuristic solution for the coverage problem (target coverage) which generates non-disjoint cover sets. Recently, a new solution heuristic is proposed by DIOP et al. [11] where cover sets are generated using greedy method. They proposed a cost metric $W$ as in (3) based on the coverage and remaining energy of sensor $s_k$ with respect to $T$ and give priority to those sensors to be part of cover set which has maximum metric value.

$$W_T^k = \sum_{z_j \in T} \frac{p_j^k}{n E_k \sum\limits_{s_l \in S} \frac{p_j^i}{E_l}}, \tag{3}$$

where $p_j^k$ equals 0 if the target $z_j$ is covered by $s_k$, and 1 otherwise. As given in (3), greedy heuristic in [11] always try to choose that sensors which is covering maximum number of sensors with more remaining energy. In this paper, we propose to advance the given method in [11] by introducing one more step in their heuristic which generates cover sets. The greedy method given by DIOP et al. [11] only generate cover set but do not try to reduce the number of total generated cover sets. To utilize the limited battery resource in optimal manner, we need to keep alive sensors for longer period. This needs that cover sets should have least possible sensors so that all the targets are covered. In order to do this, we propose to minimalize the generated cover set by heuristic in [11] to remove extra sensors. Due to this process, more cover set generated which results in increased network lifetime. The sum of remaining energies covering the critical target(s) is a theoretical upper bound on the maximum network lifetime which is achievable by any heuristic method. Therefore, the total gain in network lifetime directly depends on the remaining energies of these critical targets. In order to follow this observation, our proposed heuristic always select only one sensor in cover set which is covering critical target. The complete functionality of the proposed technique is given below (named Minimal-Heuristic).

| Minimal-Heuristic: Proposed method for target coverage |
|---|
| **Input:** |
| **S**: total number of sensor deployed in network |
| **T**: Set of targets in the same networks |
| $E_i$: initial energy level of sensor $s_i$ |
| **Output:** |
| **L**: Network Lifetime |
| **K**: Set of all cover sets |
| 01: **initialize**   L=Θ, K= Θ, $T_{uncovered}$=T |
| 02: *while* $T_{uncovered} \neq \Theta$ |
| 03:        // **till all the targets are covered** |
| 04:     *for* all $s_i \epsilon S$ *do* |
| 05:   *find* the sensor $s_i$ covering maximum uncovered targets |
| 06:     *end for* |
| 07:   *if* $s_i$ is only single sensor *then* *select* $s_i$ and **go to** step **12** |
| 08:   *end if* |
| 09:   *if* multiple sensor covering maximum number of targets *then* |
| 10:       *select* $s_i$ which has height cost as in **(3)** |
| 11:   *end if* |
| 12:       $K_p = K_p \cup s_i$ |
| 13:     *for* all target t covered by $s_i$ **do** |
| 14:            $T_{uncovered} = T_{uncovered}$ -{t} |
| 15:     *end for* |
| 16: *end while* |
| 17: minimialize generated cover set //using **Minimal_Cover()** |
| 18:    *for* all $s_i$ in minimal_$K_p$ *do* |
| 19:          $E_i = E_i$-X($K_p$) |
| 20:    *end for* |
| 21:      L=L+ X($K_p$) |
| 22:      K=K∪ {minimal _ $K_p$ } |
| 23: *return* L, K |
| 24: *exit* |

**Minimal-Heuristic**: Initially, all the targets are assumed to be uncovered (Line 1) and then we check for all the sensors and select the sensor which covers maximum uncovered targets (Line 4–5). If single sensor covering all the targets then the cover set is formed with this sensor (Line 12). If this is not the case (Line 9), then, the sensors which is covering maximum uncovered targets with highest cost (Line 10) is selected in the current sensor cover. This process is repeated until all the targets are covered (Line 13–16). In order to minimize generated cover set (Line 17), we propose another method called **Minimal_Cover** as given following:

---

**Minimal_Cover:** Minimalize Cover set $K_p$

  **Input:**    S: a set of all the sensors
              $K_p$: a Cover set
  **Output:**  minimalized_ $K_p$ // minimalized cover set
  01:  ***initialize*** minimalized_ $K_p = \Theta$
  02:  ***for*** r = length($K_p$) ***down*** to 1 ***do***
          // $s_r$. $K_p$ is $r^{\text{th}}$ **sensor in** $K_p$
  03:      ***if*** $K_p$ - ($s_r$. $K_p$) still a cover set ***then***
  04:          $K_p = K_p - s_r$. $K_p$ // ignore $s_r$. $K_p$
  05:      ***else***
  06:        minimalized_ $K_p$ = minimalized_ $K_p \cup s_r$. $K_p$
  07:      ***end if***
  08:  ***end for***
  09: ***return*** minimalized_ $K_p$
  10: ***exit***

---

Once cover set is minimalized, the remaining battery level of the sensors in this cover set is reduced by the amount X(K) (Line 18–20). Once this process is over, a cover set is generated and this process is keep repeated till a new cover set can be generated.

## 4 Simulation

Now the proposed Minimal-Heuristic and the existing methods by Cardei et al. [4], Slijepcevic and Potkonjak [5], and DIOP et al. [11] are compared. We compared all the four techniques based on the network lifetime verses number of sensor nodes, number of targets, sensing ranges, sensing area, and the execution time. Each simulation outcome is an average of 50 instances in total. We follow the radio model given in [3] where total energy consumed to send and receive an **n**-bit data packet at distance d is given in Eqs. (4) and (5) as follows:

$$E_{\text{tx}} = P \times \left( E_{\text{amp}} + \varepsilon_{fs} \times d^n \right) \tag{4}$$

$$E_{\text{rx}} = P \times E_{\text{amp}} \tag{5}$$

### 4.1 Network Lifetime Versus Sensors

In this simulation, we take fix number of targets equals 100 and keep changing sensors from 50 to 250. We consider homogenous sensor networks with fixed sensing range (150 m) and equal energy level ($E_i = 1.5$ J). The simulation outcome for this experiment is shown in Fig. 1. It is clearly evident from Fig. 1 that

**Fig. 1** Network lifetime
obtained by varying sensors

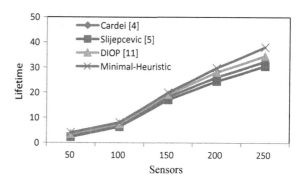

**Fig. 2** Network lifetime
obtained by varying targets

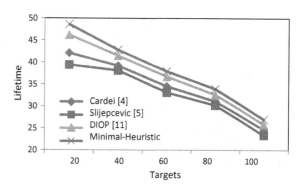

increasing the number of sensors nodes within the same sensing area results in longer network lifetime. This happens so, because keeping same sensing area, when numbers of sensors are more, then targets will be covered for longer period which result is better network lifetime. It is also evident from Fig. 1 that the proposed Minimal-Heuristic performs better than those discussed in [4, 5, 11].

## 4.2 Network Lifetime Versus Targets

Here we take fix number of sensors equals 150 and keep changing targets from 20 to 100. We consider homogenous sensor networks where each sensor has sensing range equals 150 m and equal energy level ($E_i$ = 1.5 J). As depicted in Fig. 2, the network lifetime sharply decreases by increasing the targets in the fixed sensing area. This happens so as more number of sensors required for monitoring the increased number of targets which results in lesser network lifetime. Again, proposed Minimal-Heuristic outperforms the heuristics in [4, 5, 11].

**Fig. 3** Network lifetime
obtained by varying sensing
range

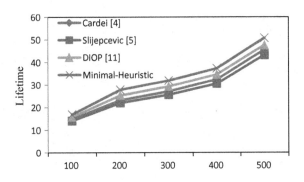

## 4.3 Network Lifetime Versus Sensing Range

Here we take fix number of targets equal 100, fixed number of sensors equal 200, and varying sensing range from 100 to 500 m. We consider homogenous sensor with equal energy level ($E_i$ = 1.5 J). It is evident from Fig. 3 that network lifetime increases when sensing range of sensor nodes increases. This happens so as that with bigger sensing range, sensor can cover more targets which results in less cardinality of cover set. Due to this, the total network lifetime also increases. As shown in previous simulations, here too proposed heuristic outperforms the existing ones.

As we have observed that under various scenarios (4.1–4.3) the network lifetime achieved by proposed Minimal-Heuristic is always better than those addressed by Cardei et al. [4], Slijepcevic and Potkonjak [5], and DIOP et al. [11].

## 5   Conclusion

In this paper, we discussed the well-known coverage problem (target coverage). To prolong the global network lifetime, we propose an optimized energy-based heuristic. Our proposed heuristic basically improves the performance of existing technique [11] by minimalizing the generated cover set. We simulated under various parameters to claim the superiority of the proposed method over some existing ones [4, 5, 11]. In future, we may extend this work for some more variants like Q-Coverage, Connected Coverage, and Variable radii Coverage.

## References

1. Akyildiz, I.F., Su, W., Sankarasubramaniam, Y., Cayirci, E.: A survey on sensor networks. IEEE Commun. Mag. **40**(8), 102–114 (2002)
2. Pujari, A.K., Mini, S., Padhi, T., Sahoo, P.: Polyhedral approach for lifetime maximization of target coverage problem. ICDCN, 1–8 (2015)

3. Cardei, M., Du, D.Z.: Improving wireless sensor network lifetime through power aware organization. ACM Wirele. Netw. **11**, 333–340 (2005)
4. Cardei, M., Thai, M.T., Li, Y., Wu, W.: Energy-efficient target coverage in wireless sensor networks. In: Proceedings of IEEE Infocom (2005)
5. Slijepcevic, S., Potkonjak, M.: Power efficient organization of wireless sensor networks. In: Proceedings of the IEEE International Conference on Communications, 472–476 (2001)
6. Gu, Y., Zhao, B.-H., Ji, Y.-S., Li, J.: Theoretical treatment of target coverage in wireless sensor networks. J. Comput. Sci Technol. **26**, 117–129 (2011)
7. Gu, Y., Li, J., Zhao, B., Ji, Y.: Target coverage problem in wireless sensor networks: a column generation based approach. In Proceedings IEEE 6th International Conference Mobile ADHOC Sensor System, 486–495 (2009)
8. Manju., Pujari, A.K.: High-Energy-First (HEF) Heuristic for Energy Efficient Target Coverage Problem. IJASUC **2**(1) (2011)
9. Chaudhary, M., Pujari, A.K.: Q-coverage problem in wireless sensor networks. ICDCN 325–330 (2009)
10. Mini, S., Udgata, S.K., Sabat, S.L.: Sensor deployment and scheduling for target coverage problem in wireless sensor networks. IEEE Sens. J. **14**(3), 636–644 (2014)
11. Diop, B., Diongue, D., THIARE, O.: Managing target coverage lifetime in wireless sensor networks with greedy set cover. Sixth International Conference on Multimedia, Computer Graphics and Broadcasting, 17–20 (2014)
12. Castaño, F., Rossi, A., Sevaux, M., Velasco, N.: A column generation approach to extend lifetime in wireless sensor networks with coverage and connectivity constraints. Comput. Oper. Res. **52**(B), 220–30 (2014)

# Effect of Black Hole Attack on MANET Reliability in DSR Routing Protocol

**Moirangthem Marjit Singh and Jyotsna Kumar Mandal**

**Abstract** The use of Mobile Ad hoc Network (MANET) for communication purpose has gained momentum recently. Hence, reliability and security aspects of MANET must be ensured. A good amount of research work on performance and security analysis of MANET are found in the literature. However, less research work on reliability analysis of MANET is currently available due to absence of any suitable method to deal with dynamic nature of MANET. In this paper, we present reliability analysis of MANET under black hole attack running DSR routing protocol. ns-2.35 simulation software is used to analyze the consequence of black hole attack on the reliability of MANET in DSR routing protocol. Simulation shows that the reliability of MANET under multiple black hole attack is lower than the reliability of MANET with single black hole attack in DSR routing protocol.

**Keywords** Mobile ad hoc network · Black hole attack · Network reliability · DSR routing protocol

## 1 Introduction

The growth of wireless networking has made Mobile Ad hoc Network (MANET) [1] an important research topic since the mid-1990s. There has been considerable use of MANET in recent times. It has become the buzzword of the day and it is an integral part of our day-to-day e-activities such as e-business that runs over digital infrastructure. MANET is made up of mobile nodes that are interconnected using

M.M. Singh (✉)
Department of Computer Science & Engineering, North Eastern Regional Institute of Science & Technology, Nirjuli 791109, Arunachal Pradesh, India
e-mail: marjitm@gmail.com

J.K. Mandal
Department of Computer Science & Engineering, University of Kalyani, Kalyani 741235, West Bengal, India
e-mail: jkm.cse@gmail.com

© Springer Nature Singapore Pte Ltd. 2018
R.K. Choudhary et al. (eds.), *Advanced Computing and Communication Technologies*, Advances in Intelligent Systems and Computing 562, https://doi.org/10.1007/978-981-10-4603-2_26

275

wireless links that forms a temporary communication network within its area of deployment. Dynamic topology change, multi-hop routing, infrastructure less, self-organizing, etc., are some of the distinctive attributes of MANET. The nodes act as a router for each other nodes in the routing process. It is useful in situations where establishment of infrastructure-based networks is not desirable. However, there exist different areas in MANET for improvement and investigation. One of the important matters is the reliability and security facet of MANET. Reliability of MANET is in nascent stage compared to that of the reliability of wired networks and it is vulnerable to security threats and attacks. A system is highly reliable; if there is insignificant probability of the system to be nonoperative at any time of its use. Hence, we need to include reliability analysis as an essential section in design and implementation of MANETs. Some research work related to reliability analysis of MANET can be found in the literature. Network reliability may be defined as the probability of victorious communication between designated nodes in the network under noted conditions of environment. Computing the probability that a group of nodes can talk/communicate to one another for a stipulated time period is the typical network reliability problem. The higher value of MANET reliability implies better service offered by MANET. Security attacks if launched successfully in the MANET can hamper normal operation of the network which also degrades network service performance affecting MANET reliability. A good amount of research work on performance and security analysis of MANET are found in the literature [2]. However, less research work on reliability analysis of MANET is currently available due to absence of any suitable method to deal with dynamic nature of MANET. Most of the researchers in MANET have overlooked the importance of reliability engineering application in MANET. Most of the works such as the one in [3] have emphasized on performance metrics such as delay, throughput, routing overhead, etc. We will make an effort to achieve filling the existing gap in reliability engineering application in MANET under security attacks through this paper. In this work, one of the especially common security attacks in MANET, i.e., the black hole attack is enforced in the MANET running DSR routing protocol and the consequence of black hole attack on the MANET reliability is analyzed.

This paper is oriented as follows. In Sect. 2, background is presented. Methodology used is stated in Sect. 3. Section 4 specifies implementation and result analysis. Conclusion is given out in Sect. 5 and references at the end.

## 2 Background

The work reported in the paper is based on the concepts of network reliability against black hole attack in MANET. Each of these concepts along with a summary of DSR routing protocol is briefly discussed in this section.

## 2.1  Network Reliability

Network reliability [4] is the probability of successful operation of a network over a stipulated time period under stated conditions of environment. It is the probability of communicating successfully between source and target/destination nodes within the network. Network reliability may also be defined as preparedness of the network to pursue activities in the event of part failures. There exists no less than one path between source and target nodes. The victorious communication probability between two designated nodes of the network is referred as two-terminal reliability. On the other hand, the victorious communication probability between the nodes of the network is referred as all-terminal reliability. MANET is more susceptible to failures and security threats due to low transmission capability, etc. Hence, MANET reliability requirements must be investigated. MANET reliability analysis is entirely different from that of traditional networks due to unique/special MANET features such as dynamic topology, multi-hop routing, self-organizing, etc. MANET is susceptible to a number of security attacks, and hence its reliability is very fragile. Accurate and adequate MANET reliability analysis can guide in MANET protocol implementation for better quality of service. Survey on MANET reliability computation methods/analysis has been done by Singh and Mandal [5]. Most of the MANET reliability computation methods developed does not consider every feature of MANET and uses some assumptions. However, the method developed by Singh et al. [6] has considered most of the MANET features and they supported the correctness of their method using ns-2.35 simulation software. Reliability analysis of MANET accompanied by varying pause times and fixed maximum speed of the mobile nodes based on logistic regression can be seen in [7].

## 2.2  Black Hole Attack

Black hole attack [8, 9] is a very frequently observed security attacks especially in MANET. Here, a false node will send a fake message to the source node claiming that it has the shortest path toward destination node. The source node will transmit the data to this false node. But the false node will drop all the packets that it receives and will never forward to the destination node. This false node is called the black hole node and this event is referred as the black hole attack. Considerable amount of literatures on black hole attacks are available.

## 2.3  Overview of DSR Routing Protocol

Dynamic Source Routing (DSR) protocol [10] is a reactive routing protocol. It finds the route/path as required and maintains them. It uses source routing that makes it loop free. Two mechanisms of DSR are route discovery and route maintenance. Each node has a cache to store latest route/path discovered. Whenever a source node wishes to transmit data to a destination node, it will search the route cache if it has a route to the destination. If the route is not available, it initiates a route discovery mechanism by broadcasting route request message (RREQ) to neighboring nodes. When the neighboring node gets the RREQ message, its cache is checked to confirm if the route is available. It responds the request of the source in unicast mode if the route is found, otherwise RREQ message is transmitted to neighboring nodes. DSR can bear relatively speedy rates of mobility.

A route error packet is generated and transmitted along the reverse path to the source in case a disconnection occurs between a node and its neighbor. All the route addresses having disconnected link addresses are deleted from the cache. A fresh route discovery process begins if an erroneous packet reaches back to the source node. The route specified at a source is dynamically modified if an alternative path to a destination is to be followed.

## 3   Methodology

The main purpose of the paper is to analyze the effect of black hole attack on the reliability of MANET running DSR routing protocol. The reliability computation method based on the concept of logistic regression developed for MANET by Singh et al. [6] is applied to examine the effect of black hole attack on the reliability of MANET in this paper. The various steps of the proposed methodology are given below.

1. Design MANET frameworks using ns-2.35 simulation tool, i.e., MANETs with node configurations 20, 30, 40, 50, 60, 70, 80, 90, 100, 110 running DSR protocol.
2. Quantify reliability of each MANET using logistic regression.

   2.1: apply the logistic function $y = 1/(1 + \exp - z)$; $-\infty < z\infty$
   2.2: $z = al + be.x_i$; $al$, $be$ are unknown variables, $x_i$ is a known variable.
   2.3: evaluate $al$ and $be$ using maximum likelihood estimation with parameters: $x_i$ (maximum speed), $t_i$ (packets transmitted), $p_i$ (packets received), $t_i - p_i$ (packets dropped).
   2.4: estimate $be$ by applying the formula: $be = (x^{\mathrm{T}}wx)^{-1}x^{\mathrm{T}}wz$, where $w_i = \mu_i(t_i - \mu_i)/t_i$, $z_i = \log\{(p_i + 1/2)/(t_i - p_i + 1/2)\}$, $\mu_i = logit^{-1}(\eta_i)$, $\eta_i = \ln(p_i/t_i - p_i)$

3. Record the reliability of every MANET obtained in step 2.
4. Impose single black hole attack in every MANETs designed in step 1.
5. Repeat step 2 to calculate MANET reliability under single black hole attack.
6. Record MANET reliability under single black hole attack obtained in step 5.

7. Impose multiple black hole attack in every MANETs designed in step 1by launching additional black hole nodes.
8. Repeat step 2 to calculate MANET reliability under multiple black hole attack.
9. Record MANET reliability under multiple black hole attack obtained in step 8.
10. Analyze the MANET reliabilities obtained in step 3, 6 and 9.

The MANET reliability values obtained are finally compared to examine the effect of black hole attack on the reliability of MANETs.

# 4 Implementation and Result Analysis

The simulation is carried out using ns-2.35 simulation software. Using the trace file generated by simulation, we obtain the number of packets transmitted, packets received and dropped packets for different maximum speeds (5, 10, 15, 20, 25 m/s) and pause times of 10 s for mobile nodes. Random way point model and constant bit rate (CBR) were used as mobility and traffic models respectively for the purpose of simulation. The packet is of 512 bytes in size. The simulation is carried on different MANETs having node configurations beginning at 20 numbers of nodes with addition of 10 nodes up to 110 nodes. Mobile nodes are deployed in an area of 4000 m × 300 m network topology. Simulation is performed for MANETs without black hole attack, with single and multiple black hole attacks running DSR routing protocol. Simulation specification and constants used for implementation are given in Table 1.

The simulation results are given in Tables 2, 3 and 4 for MANETs without black hole attack, with single and multiple black hole attacks running DSR protocol, respectively.

Figures 1, 2, and 3 show the graphical plots for reliability against number of nodes in MANETs without black hole attack, with single and multiple black hole attacks respectively. Figure 4 gives the graphical plot the compares the reliabilities of these MANETs.

| Table 1 Simulation specification | | |
|---|---|
| Network area | 4000 m × 300 m |
| No. of nodes | 20, 30, 40, 50, 60, 70, 80, 90, 100, 110 |
| Maximum speed | 5, 10, 15, 20, 25 m/s |
| Pause time | 10 s |
| MAC layer | IEEE 802.11 |
| Node mobility | Random waypoint model |
| Size (packet) | 512 bytes |
| Simulation run time | 100 s |
| Routing protocol | DSR |
| Attack type | Single and multiple black hole attacks |

**Table 2** Reliability of MANETs running DSR without black hole attack

| No. of nodes | Reliability |
| --- | --- |
| 20 | 0.050717 |
| 30 | 0.110352 |
| 40 | 0.313321 |
| 50 | 0.727015 |
| 60 | 0.770217 |
| 70 | 0.890534 |
| 80 | 0.909103 |
| 90 | 0.963101 |
| 100 | 0.992077 |
| 110 | 0.999983 |

**Table 3** Reliability of MANETs running DSR with single black hole attack

| No. of nodes | Reliability |
| --- | --- |
| 20 | 0.005502 |
| 30 | 0.007246 |
| 40 | 0.219325 |
| 50 | 0.408911 |
| 60 | 0.539152 |
| 70 | 0.393374 |
| 80 | 0.236372 |
| 90 | 0.574171 |
| 100 | 0.174454 |
| 110 | 0.699988 |

**Table 4** Reliability of MANETs running DSR with multiple black hole attack

| No. of nodes | Reliability |
| --- | --- |
| 20 | 0.005359 |
| 30 | 0.005176 |
| 40 | 0.052661 |
| 50 | 0.263508 |
| 60 | 0.385109 |
| 70 | 0.304526 |
| 80 | 0.312508 |
| 90 | 0.338947 |
| 100 | 0.253765 |
| 110 | 0.607999 |

It can be seen form Fig. 1 that the reliability value increases as the population of nodes grows in the MANET without black hole attack. The reliability of MANET either increases or decreases showing inconsistent values upon introduction of black hole attack in the MANET as shown in Figs. 2 and 3. However, the reliability of MANET is lowest under multiple black hole attacks except at few points as seen

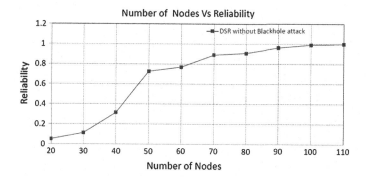

**Fig. 1** Graphical plot for reliability of MANET running DSR without black hole attack

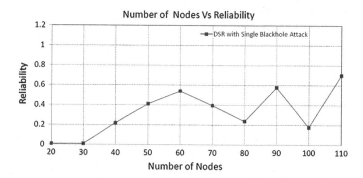

**Fig. 2** Graphical plot for reliability of MANET running DSR with single black hole attack

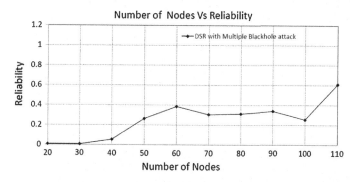

**Fig. 3** Graphical plot for reliability of MANET running DSR with multiple black hole attack

from Fig. 4. The reliability of MANET with multiple black hole attacks is lower than the reliability of MANET with single black hole attack running DSR protocol in general as evident from simulation result and graphical plots. However, MANET reliability increases as the number of nodes grows in all MANET configurations.

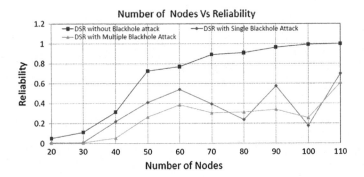

**Fig. 4** Graphical plot that compares the reliability of Figs. 1, 2 and 3

Further, the MANET reliability is likely to decrease with a rise in population of black hole nodes of the network. This is because black hole nodes keep on dropping the packets and hence the reliability value drops down.

## 5 Conclusion

The paper has furnished the outcome of black hole attack on the reliability of MANETs running DSR routing protocol with single and multiple black hole attacks. It can be clearly ascertained from the simulation results that the reliability of MANETs with multiple black hole attack is lower than the reliability of MANETs with single black hole attack in general. Increasing the quantity of black hole nodes in the MANET will lower MANET reliability. The scope of the research may be extended further by considering other network security attacks and threats.

## References

1. Sarkar, S.K., Basabaraju, T.G., Puttamadappa, C.: Adhoc Mobile Wireless Networks: Principles, Protocols and Applications, 2nd edn. CRC Press (2013)
2. Khan, M.S., Jadoon, Q.K., Khan, M.I.: A comparative performance analysis of MANET routing protocols under security attacks. In: Kim K.J., Wattanapongsakorn (eds) Mobile and Wireless Technology 2015. Lecture Notes in Electrical Engineering, vol. 310, pp. 137–145. doi:10.1007/978-3-662-47669-7_16. Springer (2015)
3. Chang, J.-M., Tsou, P.-C., Woungang, I., Chao, H.-C., Lai, C.-F.: Defending against collaborative attacks by malicious nodes in MANETs: a cooperative bait detection approach. IEEE Syst. J. Vol. 9(1), March 2015, pp. 65–75, (2015)
4. Shooman, M.L.: Reliability of Computer Systems and Networks, 1st edn. Wiley, New York (2002)

5. Singh, M.M., Mandal, J.K.: Reliability analysis of mobile adhoc network. In: 7th IEEE International Conference on Computational Intelligence and Communication Network, pp. 161–164. 12–14 Dec 2015, Jabalpur, India (2015)
6. Singh, M.M., Baruah, M., Mandal, J.K.: Reliability computation of mobile adhoc network using logistic regression. In Proceedings of 11th IEEE-IFIP International Conference on Wireless and Optical Communications Networks, Vijayawada, India. doi:10.1109/WOCN.2014.6923060 (2014)
7. Singh, M.M., Mandal, J.K.: Logistic regression based reliability analysis for mobile adhoc network with fixed maximum speed and varying pause times. J. Sci. Ind. Res.Vol. **76**, February 2017, pp. 81–84, (2017)
8. Tseng, H., Chou, L.-D., Chao, H.-C.: A survey of black hole attacks in wireless mobile adhoc networks. Hum.-Cent. Comput. Inf. Sci. (Springer Open Journal) **1**(4), 1–16 (2011)
9. Bar, R.K., Mandal, J.K., Singh, M.M.: QoS of MANet through trust based AODV routing protocol by exclusion of black hole attack. Elsevier Procedia Technol. Vol.**10**, pp. 530–537. doi:10.1016/j.protcy.2013.12.392 (2013) [Elsevier]
10. Johnson, D.B., Maltz, D.A.: Dynamic source routing in adhoc networks. In: Imielinski, T., Korth, F.H. (eds) Mobile Computing, vol. 353. The Kluwer International Series in Engineering and Computer Science, Springer, US, pp. 153–181. doi:10.1007/978-00585-29603-6_5 (1996)

# Part X
# Microelectronics and Antenna Design

# Noise Performance and Design Optimization of a Piezoresistive MEMS Accelerometer Used in a Strapdown Medical Diagnostic System

**Sonali Biswas and Anup Kumar Gogoi**

**Abstract** This paper presents the noise performance and design optimization of a silicon piezoresistive MEMS accelerometer with a frequency range of (0.1–25 Hz) and a dynamic range of $\pm 2g$ to be used in a strapdown physiological tremor diagnostic system. The MEMS accelerometer designed is based on the simple mass spring damper system and simulated using Finite element method-based software COMSOL 4.3. Here the proofmass is a quad surrounded by four flexures; two on either side and the entire structure is supported by a fixed frame. For sensing stress at maximum points total no. of eight p-doped piezoresistors are implanted; four at the junction of the mass and flexures and the other four at the flexure and fixed frame junction. The noise spectrum has been obtained from the fundamental equation of the system and has been plotted for different quality factors. The accelerometer noise for the designed device with desired damping ratio of 0.8 and Quality factor $Q = 0.6$ is obtained as $8.1\ \mu m/s^2/\sqrt{Hz}$. In order to increase the signal-to-noise ratio, the option was to increase the mass and quality factor and reduce the resonating frequency. Proofmass increase counters the miniaturization, high $Q$ results in excessive ringing effect also it requires enough dynamic range. Further, if the resonating frequency is reduced, it may introduce nonlinear phase into the system. Hence for lowering the noise floor to achieve higher performance in terms of sensitivity, optimizations of parameters are important. Also in order to enhance the performance of the device, a noise reduction scheme has been proposed.

**Keywords** Piezoresistive · MEMS accelerometer · Proofmass

S. Biswas (✉) · A.K. Gogoi
Department of Electronics and Electrical Engineering, Indian Institute of Technology, Guwahati 781039, India
e-mail: b.sonali@iitg.ernet.in

A.K. Gogoi
e-mail: akg@iitg.ernet.in

© Springer Nature Singapore Pte Ltd. 2018
R.K. Choudhary et al. (eds.), *Advanced Computing and Communication Technologies*, Advances in Intelligent Systems and Computing 562, https://doi.org/10.1007/978-981-10-4603-2_27

# 1  Introduction

Micro-accelerometers and other micro-sensors have been a boon in various fields and the cost has been reduced because of batch fabrication. However, the smaller the device the inferior is the signal-to-noise ratio [1]. With an objective of increasing the performance of the system, an attempt has been made to study the sources of noise and thereby optimize the design parameters, considering the challenges. The MEMS (microelectromechanical system) accelerometer designed is of the piezoresistive type and the desired properties include high sensitivity and resolution, maximum operating range, wide frequency response, good linearity, low cross axis sensitivity, and high signal-to-noise ratio. The first accelerometer work [2] was followed by various acceleration sensing by different types of piezoresistive accelerometers by many researchers all having its merits and demerits. The piezoresistive MEMS accelerometer designed in this work is based on the basic mass spring damper system [3, 4]. The structure with the design configuration has been simulated using finite element method-based software COMSOL 4.3. There are several reasons for choosing piezoresistive method of transduction, mainly because of simplicity in structure, better sensing, easy fabrication, and ruggedness. Throughout the time span, various MEMS piezoresistive accelerometers have been developed for several applications in different emerging fields' w.r.t performance. However, miniaturization itself is a challenge which makes signal-to-noise ratio improvement a must for high performance. This paper primarily focuses on the mechanical thermal noise in the MEMS-based strapdown tremor diagnostic system in Sects. 3–5. In order to enhance the sensor performance, a noise reduction scheme has been proposed using a sigma delta modulator in Sect. 6. The dissipation mechanism which has impact upon the noise performance is given in Sect. 7 with the results and discussion in Sect. 8.

# 2  Noise Sources

Noise is an unwanted signal and sometimes high noise floor can make measurement of the signal of interest very difficult. The Intrinsic non-deterministic noise which originates from the device itself can neither be neglected nor avoided by shielding. Noise being a random process, for measurement purpose we take the power spectral density function, which gives the magnitude of the random signal squared over a range of frequencies. The fundamental noise mechanisms that potentially limit the performance of the piezoresistive MEMS sensors are mechanical thermal noise, Johnson noise, Hooge noise or $1/f$ noise, and shot noise. Miniaturization is attractive for many applications but the small moving parts are especially suscep-tible to mechanical noise resulting from molecular agitation. Any molecular agi-tation even through solid structure like springs and support can cause random motion of an object which is called the Brownian motion. The mechanical thermal noise depends on the temperature and the magnitude of mechanical damping. Johnson noise is independent of frequency and it occurs due to thermal energy in a resistor. The expression of the thermal noise power spectral density in ($V^2$/Hz) as units have been reported in [5, 6] and is given by Eq. (1)

$$S_J = 4K_B RT,\qquad(1)$$

where $K_B$ is the Boltzmann constant, $R$ is the resistance, and $T$ is the temperature in Kelvin. Electrical thermal noise is caused by the agitation of the charge carriers by thermal lattice vibrations and is present regardless of bias voltage. But higher temperature induces more agitation of the carriers; hence, the Johnson noise is temperature dependent. Moreover as the lattice vibrations are random and not related to any single time constant, therefore, Johnson noise is frequency independent. The mechanical thermal noise depends on the temperature and the magnitude of mechanical damping. Mechanical thermal noise is analogous to electrical thermal noise. The fluctuation-dissipation theorem says that there must be a fluctuation force to maintain the energy balance and the thermal equilibrium, if there exist mechanical damping in any dissipative mechanism [6]. The mechanical thermal noise equation analogous to the above equation is given by Eq. (2) and their unit is in ($N^2$/Hz).

$$S_m = 4K_B R_m T\qquad(2)$$

Here $R_m$ is the equivalent mechanical resistance of the sensor.

Low Frequency noise also known as $1/f$ noise or Hooge is a frequency dependent non-equilibrium noise. Though the main reasons are the fluctuation in mobility and carriers [7], still it is an active research area.

In a semiconductor, when charge carriers cross a potential barrier independently and randomly, fluctuations which occur in the average current give rise to shot noise which is a non-equilibrium noise. The shot noise power spectral density in ($A^2$/Hz) as units is given by Eq. (3)

$$S_I = 2qI\qquad(3)$$

where $q$ is the electron charge and $I$ is the current. The shot noise is frequency independent. In this paper, the mechanical thermal noise for the designed piezoresistive accelerometer has been derived under thermal equilibrium and by adding a force generator alongside the damper. By solving the accelerometer for noise response we get an idea of the various dependent design parameters we need to consider to increase the signal-to-noise ratio. The dissipation mechanism has been studied which gives an idea to choose appropriate damping mechanism and the damping ratio. The low frequency dominant noise has been reduced by suitable doping concentration thus enhancing the resolution and sensitivity. Thus taking into consideration the various design and process parameters and the necessary tradeoffs, a sigma delta noise reduction model has been proposed for the high performance.

## 3 Strapdown Tremor Diagnostic System and Noise

Tremor occurs in patients suffering from neurodegenerative diseases. The tremor occuring in patients which can be sensed by a triaxial accelerometer can be corrupted by noise signals. Noise may occur due to several reasons. During signal pick up and processing noise may occur resulting from the instrumentation error, noisy environment, low frequency noise due to movement artifacts, etc. Hence, minimization of noise becomes important for proper diagnosis of the tremor occuring in such patients. Hence in order to have higher signal-to-noise ratio, our objective is to design a strapdown tremor diagnostic system with minimum noise floor.

## 4 Device Configuration

The microaccelerometer designed has a silicon square proof mass with four flexures or beam, two on either side of the proof mass. The entire structure is surrounded by a fixed frame. Eight p-doped single crystal silicon piezoresistors are implanted, four on the junction of the beam and proofmass and the other four on the junction of the beam and fixed frame. The chosen geometric dimensions: Proofmass 3200 μm × 3200 μm × 250 μm, Flexures-1000 μm × 250 μm × 20 μm, Frame-5200 μm × 230 μm × 250 μm, and piezoresistors 100 μm × 25 μm × 2 μm.The structure has been simulated using finite element tool COMSOL 4.3. Here the effective spring constant is taken for parallel beams which is given by four times the spring constant for each. The design parameters obtained has been shown in Table 1. The signal pick up circuit is the Wheatstone bridge where all the eight piezoresistors form the arms of the bridge, this helps in reducing the cross axis sensitivity [8]. The voltage output is directly proportional to the applied acceleration.

**Table 1** The device design parameters

| Design parameter | Units | Symbol | Value |
|---|---|---|---|
| Proofmass weight | kg | $m$ | $5.96 \times 10^{-6}$ |
| Effective spring constant | N/m | $K$ | 1352 |
| Resonating frequency | rad/s | $\omega_0$ | 15,061 |
| Regular frequency | Hz | $f$ | 2397 |
| Damping coefficient | N/(m/s) | $\gamma$ | 0.143 |
| Quality factor | – | $Q$ | 0.625 |
| Damping ratio | – | $\xi$ | 0.8 |

## 5 Noise in the Accelerometer

The performance and the design optimization of the MEMS accelerometer can be made from the accelerometer noise [1]. The fluctuation-dissipation theorem gives a statistically balancing equation between the energy lost by the system through damping and the energy brought into the system by the equivalent noise forces. For the overall noise analysis of the structure we use the equipartition and Nyquist theorems. Considering thermal equilibrium, the equation of motion is Eq. (4).

$$m\frac{d^2x}{dt^2} + \gamma_d\frac{dx}{dt} + kx = F_n \tag{4}$$

The above equation in terms of velocity $v$ taking $v = dx/dt = sx$ can be written as

$$msv + \gamma_d sx + kv/s = F_n \tag{5}$$

$$msv + \gamma_d v + kv/s = F_n \tag{6}$$

The mean square velocity due to the noise generator $F_n$ is given by Eq. (7)

$$v^2 = \frac{F_n^2}{\gamma_d^2 + (\omega m - k/\omega)^2} \tag{7}$$

We know the natural undamped frequency is given by

$$\omega_0 = \sqrt{\frac{k}{m}} \tag{8}$$

and the ratio of the peak amplitude to the amplitude at low frequency is referred to as $Q$, the quality factor. The sharpness of the peak is required for high resolution.

$$Q = \frac{\omega_0 m}{\gamma} \tag{9}$$

Equation (7) can be rewritten in terms of the $\omega_0$ and $Q$ as in Eq. (10)

$$v^2 = \frac{1}{\gamma^2}\frac{F_n^2}{1 + Q^2\left(\frac{\omega}{\omega_0} - \frac{\omega_0}{\omega}\right)^2} \tag{10}$$

The kinetic energy stored therefore is given by Eq. (11)

$$\overline{v_n^2} = \frac{1}{4\pi\gamma} \int\limits_0^\infty \frac{F_n^2 Q d\left(\frac{f}{f_0}\right)}{1 + Q^2\left(\frac{f}{f_0} - \frac{f_0}{f}\right)^2} \tag{11}$$

The spectral density of the fluctuating noise force related to damping is given by Eq. (12)

$$\overline{F_n^2} = 4K_B T\gamma \tag{12}$$

From Eqs. (4) and (12) we obtain the value of displacement $x$ and expressed as noise spectrum in Eq. (13)

$$\overline{x_n^2} = \frac{4K_B T\gamma}{\gamma^2\omega^2 + (k - m\omega^2)^2} \tag{13}$$

The accelerometer used to detect motion below the resonance frequency as given in Eq. (14), where

$$\overline{x_n^2} = \frac{4K_B T\gamma}{k^2} \tag{14}$$

The rms value of the displacement noise thus obtained in Eq. (15) has been plotted for various quality factors as shown in Fig. 1.

$$\langle x_n \rangle = \frac{\sqrt{4K_B T\gamma}}{k} \tag{15}$$

**Fig. 1** Noise response of the accelerometer for different quality factor

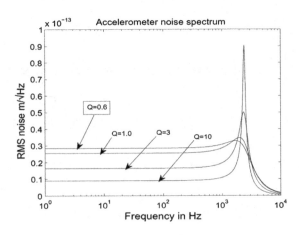

In the microaccelerometer when the input is acceleration, the mass $M$ converts the acceleration into a force that is converted to displacement by the spring with its spring constant $k$. We know by Hooke's law the force $F$ needed to extend or compress a spring by some distance $x$ is proportional to the distance i.e. $F = -kx$, where $k$ is the stiffness constant. Also from Newton's second law we know Force $F$ is given by the product of mass and acceleration.

Therefore, the expression for the rms value of the acceleration $a_n$ is obtained by equating the forces as mentioned above and we get Eq. (16).

$$\langle a_n \rangle = \frac{\sqrt{4K_B T \gamma}}{m} \tag{16}$$

For the designed piezoresistive MEMS accelerometer, the accelerometer noise obtained from Eq. (16) is 8.1 $\mu m/s^2/\sqrt{Hz}$. The noise equation (16) can be rewritten in terms of the Quality factor as in Eq. (17)

$$\langle a_n \rangle = \sqrt{\frac{4K_B T \omega_0}{mQ}}. \tag{17}$$

## 6  Proposed Scheme for Noise Reduction

The accelerometer has piezoresistors at the edges of the suspension beams in order to obtain maximum stress. When the device is moving with an acceleration perpendicular to the chip plane, the inertial force of the mass forces the beam to bend and causes stress in the beam which results in change in resistance. The signal to be picked up is done by connecting a Wheatstone bridge where the arms have the piezoresistors. Thus, the output voltage obtained is directly proportional to the accleration. This analog output obtained from the MEMS accelerometer with the Wheatstone bridge circuit is then taken and fed to a sigma delta converter. First, the sigma delta converter converts the analog signal into digital with a very low resolution analog to digital converter (ADC). The effective resolution is increased by using oversampling techniques in conjunction with noise shaping and using digital filter. The digital filter is used to attenuate signals and noise that are outside the band of interest followed by decimation. During decimation the data rate is reduced from the oversampling rate without losing any necessary information. The proposed scheme for noise reduction in the accelerometer has been given as in Fig. 2

As shown in Fig. 3 in the sigma delta modulator, first the analog input signal is fed via a summing junction. The signal is then passed through the integrator so that there can be a low-pass filtering effect and we can focus on our desired frequency up to 25 Hz and then this signal of desired band is fed to a comparator. The comparator acts as a one bit quantizer. The output of the comparator then becomes the input to the summing junction via a single bit digital-to-analog converter

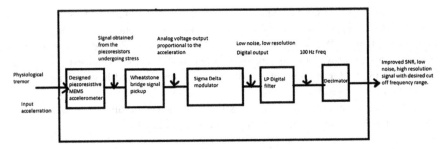

**Fig. 2** Proposed noise reduction scheme with desired frequency range

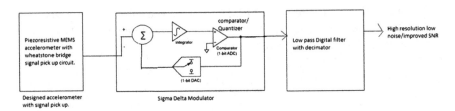

**Fig. 3** Sigma delta modulator for noise reduction

(DAC) feedback arrangement. The comparator output it also passes through the digital filter and decimator for high resolution. The feedback loop forces the average of the signal entering from the DAC to be equal to the signal fed.

The key feature of the converter is that it is of low cost but it provides high dynamic range and flexibility in handling low bandwidth input signals for analog to digital conversion.

## 7 Dissipation Mechanism

In order to evaluate the mechanical thermal noise, we should take into consideration the sources of dissipation or energy loss. Dissipation is termed as the mechanism that allows energy to escape from the orderly motion of the sensor. These include mechanical damping in the spring and support, viscous air damping, electrical leakage, magnetic eddy current damping, etc. Damping determines the thermal fluctuations, hence in order to have an improvement in the signal-to-noise ratio the device design should have an effective and controlled damping mechanism. For the piezoresistive sensor having proofmass and the top and bottom cover lids, damping is determined both by the spacing between them and the viscosity of the fluid used. This type of damping occurs when the proofmass is pushed toward a fixed microstructure with the fluid in it. Squeeze film damping dominates the dissipation mechanism for gaps of several micrometers. In the present case, the piezoresistive

| Table 2 The design parameters obtained for different $Q$ value | Quality factor | Damping ratio | Damping coefficient |
|---|---|---|---|
| | 0.6 | 0.8 | 0.143 |
| | 1 | 0.5 | 0.089 |
| | 3 | 0.16 | 0.028 |
| | 10 | 0.05 | 0.0089 |

accelerometer senses the acceleration in the vertical $z$-axis therefore squeeze film damping is considered. The damping force arises from the squeeze film effect of the proofmass and the air film trapped in the gap between the mass and the encapsulation. A large area to gap ratio results in higher squeeze number which results in greater damping. A small damping coefficient translates into lower energy loss, smaller damping ratio, and hence higher $Q$. We can improve the quality factors by reducing the operating pressure, improving the surface roughness, thermal annealing, and by modifying the boundary conditions.

# 8 Results and Discussions

The Damping ratio is a very important design parameter for the MEMS accelerometer and its value can be determined from the geometric dimension, pressure, density and fluid viscosity. It is very important that the design should be such that the damping ratio is neither very much lesser nor very much greater than unity. If we choose $\xi \ll 1$ then the structure may collapse, and if $\xi \gg 1$ then the settling time will be too large in other words it will give very slow response. In the present work the MEMS accelerometer has been designed to achieve a damping of 0.8. The design parameters obtained for different $Q$ value have been tabulated as shown in Table 2.

For the designed Piezoresistive MEMS accelerometer, the displacement rms noise obtained is $0.395 \times 10^{-13}$ m/$\sqrt{Hz}$ and the displacement noise spectrum has been plotted in Fig. 1 for various Q values. The accelerometer noise given by Eq. (16) is obtained as 8.1 $\mu$m/s$^2$/$\sqrt{Hz}$.

# 9 Conclusions

The micro accelerometer design aspects including the dissipation mechanism, signal processing circuit, temperature, etc., have an impact upon the noise performance. Hence, careful design optimization has to be done to minimize noise which includes increasing the quality factor, reducing the resonating frequency, and increasing the mass. The system must have enough dynamic range therefore the quality factor cannot be very high. A very high $Q$ can give rise to very large out of band oscillations, and introduces ringing effect. Increasing mass increases sensitivity on one hand and a reduction in resonating frequency on the other hand. The bandwidth of the accelerometer has to be traded off with its sensitivity. The resonating frequency though can be decreased, it must be taken care so that it does not fall within the band of operating frequency. Hence, mass configuration should be wisely chosen by considering the trade offs. Also the device should have optimized damping by choosing appropriate damping mechanism. In order to lower the noise floor and improving the resolution, the signal has to undergo low-pass filtering and up to 25 Hz the necessary noise reduction can be done using the proposed scheme for the tremor diagnostic system.

# References

1. Gabrielson, T.B.: Mechanical-thermal noise in micromachined acoustic and vibration sensors. IEEE Trans. Electron Devices **40**(5), 903–909 (1993)
2. Roylance, L.M., Angel, J.B.: A batch fabricated silicon accelerometer. IEEE Trans Electron Devices **26**, 1911–1917 (1979)
3. Kal, S., et al.: CMOS compatible bulk micromachined silicon piezoresistive accelerometer with low off-axis sensitivity. J. Microelectron. **37**, 22–30 (2006)
4. Ravi Sankar, A., Grace Jency, J., Das, S.: Design, fabrication and testing of a high performance silicon piezoresistive Z-axis accelerometer with proof mass-edge-aligned-flexures. J. Microsys. Technol. **18**, 9–23 (2012)
5. Johnson, J.B.: Thermal agitation of electricity in conductors. Phys. Rev. **32**, 97–109 (1928)
6. Nyquist, H.: Thermal agitation of electric charge in conductors. Phys. Rev. **32**, 110–113 (1928)
7. Hooge, F.N.: 1/f Noise is no surface effect. Phys. Lett. A **29**, 139–140 (1969)
8. Biswas, S., Gogoi, A.K.: Design issues of piezoresistive MEMS accelerometer for an application specific medical diagnostic system. **33**(1), 11–16 (2016)

# Current Mode Universal Filter Realization Employing CCIII and OTA-A Minimal Realization

**Tajinder Singh Arora**

**Abstract** Being extensively used and widely explored, continuous time filters hold a prominent place in the field of analog circuits. The paper introduces a universal filter employing third-generation current conveyor and operational transconductance amplifier with minimum passive components. The proposed design works in current mode and uses grounded passive components only, making it a better proposition for integrated circuit implementation. The working of the circuit has been tested at high frequency with electronic tunability of quality factor. Frequency response of all five basic filtering functions along with the sensitivity analysis has been included to verify the theoretical results.

**Keywords** Universal filter · Third-generation current conveyor · Current mode circuits · Active filters

## 1 Introduction

Versatility and scope of digital signal processing has been continuously dominating the world, but still there are many areas of application where analog signal processing cannot be replaced. There are various application of analog circuits such as amplification, wave form generation (oscillators), continuous time filtering, etc. [1].

Continuous time filtering is one of most widely explored area of analog signal processing and numerous papers on the same has been made available in the literature employing different active devices such as operational transconductance amplifier (OTA), current differencing buffered amplifier (CDBA), current conveyors (CC), etc., names a few [2–4].

There are various ways of characterizing a continuous filter. One way is based on the number of input and output terminals it has, i.e., SISO (single-input-single-output), SIMO (single-input-multiple-output), MIMO

T.S. Arora (✉)
Maharaja Surajmal Institute of Technology, Janak Puri, New Delhi, India
e-mail: tajarora@msit.in

© Springer Nature Singapore Pte Ltd. 2018
R.K. Choudhary et al. (eds.), *Advanced Computing and Communication Technologies*, Advances in Intelligent Systems and Computing 562, https://doi.org/10.1007/978-981-10-4603-2_28

(multiple-input-multiple-output), MISO (multiple-input-single-output). Another way of characterization is on the basis of mode of operation such as current mode (CM), voltage mode (VM), transconductance mode (TC), and trans-resistance mode (TR). The availability of the outputs can also be a naming criterion, such as if the filter is capable of delivering all the five basic responses, i.e., low pass (LP), high pass (HP), band pass (BP), band reject (BR), and all pass (AP), it is termed as universal filter. If the filter has a capacity of giving only few responses, but not all five, then it is named as multifunction filter.

Variety of universal filter realization using third-generation current conveyor (CCIII) and operational transconductance amplifier (OTA) are available in literature [5–9] and references cited therein. Significant comparison among the proposed work and the earlier published work is given in Table 1. It is evident from the tabulation that the proposed design fulfills all the constraints with minimum number of active devices.

This manuscript proposes a universal filter that employs 2 CCIII and 1 OTA as an active device. The designed filter not only employs the minimum passive components but also ensures that all are grounded in nature, making it a better design. Explicit current output is available for all the five basic filtering functions, i.e., low pass (LP), high pass (HP), band pass (BP), band reject (BR), and all pass (AP). The circuit works on high frequency and it has electronic tunability of $Q_0$. The following section gives the introduction of the used active devices i.e., CCIII and OTA. Section 3 provides the circuit diagram along with the transfer functions of the proposed current mode universal filter. Sensitivity analysis has been represented in Sect. 4. Section 5 demonstrates the simulation results of the designed filter carried out with the help of PSPICE. The concluding remarks have been given at last.

**Table 1** Comparison with the other existing work

| S. No. | Number of active devices used | Whether all grounded capacitor employed | Whether all grounded passive components | All 5 type of frequency response possible |
|---|---|---|---|---|
| [5] | 3 OTA + 1 CCII = 4 | Yes | Yes | Yes |
| [6] | 3 OTA + 1 CCII = 4 | Yes | Yes | Yes |
| [7] | 1 OTA + 4 CCCII = 5 | Yes | Yes | Yes |
| [8] | 3 OTA + 1 CCII = 4 | Yes | Yes | Yes |
| [9] | 4 OTA + 1 CCII = 5 | Yes | Yes | Yes |
| Proposed circuit | 1 OTA + 2 CCIII = 3 | Yes | Yes | Yes |

## 2  Introduction to CCIII and OTA

A third-generation current conveyor (CCIII), symbolically shown in Fig. 1, is characterized by

$$\begin{bmatrix} I_y \\ V_x \\ I_{z\pm} \end{bmatrix} = \begin{bmatrix} 0 & -\alpha & 0 \\ \beta & 0 & 0 \\ 0 & \pm\gamma & 0 \end{bmatrix} \begin{bmatrix} V_y \\ I_x \\ V_{z\pm} \end{bmatrix} \tag{1}$$

where $\alpha$, $\beta$ and $\gamma$ represent non-ideal port transfer ratios of $X$, $Y$, and $Z$ terminals, respectively, and ideally $\alpha = \beta = \gamma = 1$.

The operational transconductance amplifier (OTA), symbolically shown in Fig. 2, is characterized by

$$I_o = \pm g_m (V_+ - V_-) \tag{2}$$

The transconductance factor is defined by $g_m$ and it is controlled by the bias current inside the circuitry of the OTA. The polarity symbol + or − used before the transconductance parameter represents the direction of the current at the output.

**Fig. 1** Device symbol of CCIII

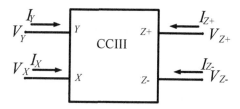

**Fig. 2** Device symbol of OTA

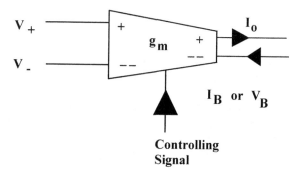

## 3 Proposed Circuit

The proposed explicit current output universal filter is given in Fig. 3. The filter circuit has three outputs namely $I_{HP}$, $I_{BP}$, $I_{LP}$, and one input as $I_{IN}$. By providing the input current to the input terminal $I_{IN}$, one can simultaneously obtain high-pass filter response ($I_{HP}$), band-pass filter response ($I_{BP}$), and low-pass filter response ($I_{LP}$). All the five basic filter responses are given by (3)–(7) along with common denominator polynomial in (8). The quality factor $Q_0$ and natural angular frequency $\omega_0$ are also given in (9).

$$\frac{I_{LP}}{I_{in}} = \frac{\left(\frac{g_{m2}}{R_1 C_1 C_2}\right)}{D(s)} \tag{3}$$

$$\frac{I_{BP}}{I_{in}} = -\frac{\frac{s}{R_1 C_1}}{D(s)} \tag{4}$$

$$\frac{I_{HP}}{I_{in}} = \frac{s^2}{D(s)} \tag{5}$$

$$\frac{I_{BR}}{I_{in}} = \frac{s^2 + \left(\frac{g_{m2}}{R_1 C_1 C_2}\right)}{D(s)} \tag{6}$$

$$\frac{I_{AP}}{I_{in}} = \frac{s^2 - \frac{s}{R_1 C_1} + \left(\frac{g_{m2}}{R_1 C_1 C_2}\right)}{D(s)}, \tag{7}$$

where

$$D(s) = s^2 + \frac{s}{R_1 C_1} + \frac{g_{m2}}{R_1 C_1 C_2} \tag{8}$$

$$\omega_0 = \sqrt{\frac{g_{m2}}{R_1 C_1 C_2}} \quad \text{and} \quad Q_0 = \sqrt{\frac{R_1 C_1 g_{m2}}{C_2}} \tag{9}$$

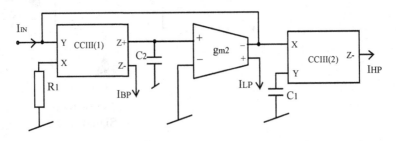

**Fig. 3** Proposed current mode universal filter

Band reject current mode response, i.e., ($I_{BR}$) is obtained by simply summing the currents available at $I_{HP}$ and $I_{LP}$ port, similarly summation of all three explicit current outputs gives the all pass response. The availability of all five basic responses makes the designed circuit a universal filter.

## 4 Sensitivity Analysis

The sensitivity analysis of the proposed configuration has been carried out and shown below. These sensitivities are well under acceptable limits.

$$S_{R_1}^{\omega_0} = S_{C_1}^{\omega_0} = S_{C_2}^{\omega_0} = -\frac{1}{2} \tag{10}$$

$$S_{g_{m2}}^{\omega_0} = \frac{1}{2} \tag{11}$$

$$S_{R_1}^{Q_0} = S_{C_1}^{Q_0} = S_{g_{m2}}^{Q_0} = \frac{1}{2} \tag{12}$$

$$S_{C_2}^{Q_0} = -\frac{1}{2} \tag{13}$$

## 5 Simulation Results

The PSPICE simulation software has been used to test and verify the designed current mode filter circuit. The MOS realizations of the CCIII [10] and OTA [11] are redrawn and shown in Figs. 4 and 5, respectively. All the MOS transistors are operating in saturation mode. The aspect ratios of all the MOS transistors, utilized in Fig. 4, are given in Table 2 and the aspect ratios for OTA are given in Table 3. The 0.18 μm CMOS model parameters used for the PSPICE simulations are taken from [12]. The selected supply voltages are $V_{DD} = -V_{SS} = 1.45$ V for the CCIII and $V_{DD} = -V_{SS} = 1.0$ V for OTA. The bias current is taken as 2.8 μA for OTA to get the values of $g_{m2}$ as 31.57 μA/V. The $\omega_0$ chosen for the design is 1 MHz and the passive component values selected are $R_1 = 17.1$ K, $C_1 = C_2 = 6.8$ pF. Simulation results of the proposed current mode filter for LP, BP, HP, and BR are shown in Fig. 6. Continuous lines depict the simulated outputs, whereas dashed lines represent the ideal or theoretical outputs of the derived filter configuration. One can see in Fig. 6 that the simulated and ideal responses are in good agreement to each other. For AP response, the magnitude as well as phase has been presented in Fig. 7. While keeping the value of $\omega_0$ fixed and taking different combination values of $g_{m2}, R_1, C_1$, and $C_2$, tunability of BP response has been provided in Fig. 8.

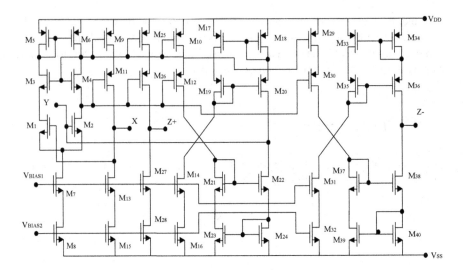

**Fig. 4** Third-generation current conveyor (CCIII) [10]

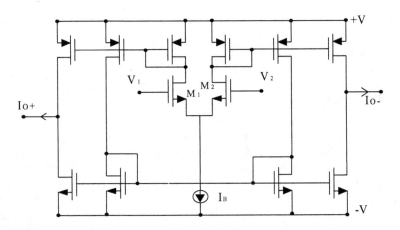

**Fig. 5** CMOS realization of the OTA [11]

**Table 2** Aspect ratios of the MOSFETs of CCIII of Fig. 4

| CMOS transistors | $W$ (μm) | $L$ (μm) |
|---|---|---|
| M1–M4 | 5.714 | 0.43 |
| M5, M6, M7, M8, M13–M16, M21–M24, M27, M28, M31, M32, M37–M40 | 9.285 | 0.43 |
| M9–M12, M17–M20, M25, M26, M29, M30, M33–M36 | 18.571 | 0.43 |

**Table 3** Aspect ratios of OTA

| MOSFET | $W$ ($\mu$m) | $L$ ($\mu$m) |
|---|---|---|
| M1, M2 | 5.76 | 0.72 |
| M3, M4, M5, M6, M7, M8 | 2.16 | 0.72 |
| M9, M10, M11, M12 | 1.44 | 0.72 |

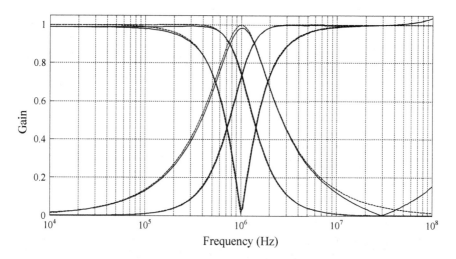

**Fig. 6** Frequency response of the proposed universal filter circuit

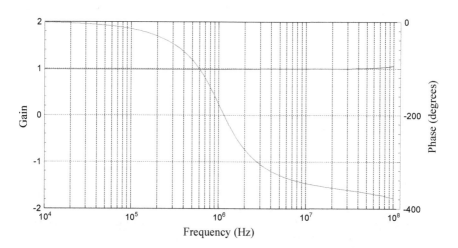

**Fig. 7** Gain and phase response of the proposed filter circuit—as an all pass filter

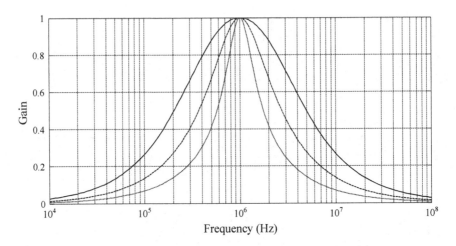

**Fig. 8** Tunability of the $Q_0$-factor for the band-pass filter circuit

## 6   Conclusion

An explicit output current mode universal filter, employing CCIII and OTA as active devices, has been proposed. With minimum number of passive components and workability at high frequencies, this circuit involves all the desired features. All the engaged passive components are grounded in nature, ensuring easier integrated circuit implementation. Use of only three passive components and electronic tunability of $Q_0$ makes it appreciable. The circuit has been successfully tested and simulated using PSPICE software. All the five output can be achieved without any matching constraints.

## References

1. Senani, R., Singh, A.K., Singh, V.K.: Current Feedback Operational Amplifiers and Their Applications. Springer Science & Business Media (2013)
2. Senani, R., Bhaskar, D.R., Singh, A.K.: Current Conveyors: Variants, Applications and Hardware Implementations. Springer, berlin (2014)
3. Arora, T.S., Rana, U.: Multifunction filter employing current differencing buffered amplifier. Circuits Syst. **7**(05), 543 (2016)
4. Hwang, Y.-S., Chen, J.-J., Li, J.-P.: New current-mode all-pole and elliptic filters employing current conveyors. Electr. Eng. **89**(6), 457–459 (2007)
5. Tsukutani, T., Edasaki, S., Sumi, Y., Fukui, Y.: Current-mode universal biquad filter using OTAs and DO-CCII. Frequenz **60**(11–12), 237–240 (2006)
6. Biolek, D., Siripruchyanun, M., Jaikla, W.: CCII and OTA based current–mode universal biquadratic filter. In: The Sixth PSU Engineering Conference, pp. 238–241 (2008)
7. Siripruchyanun, M., Jaikla, W.: Cascadable current-mode biquad filter and quadrature oscillator using DO-CCCIIs and OTA. Circuits, Syst. Sig. Proc. **28**(1), 99–110 (2009)

8. Tsukutani, T., Sumi, Y., Fukui, Y.: Novel current-mode biquad filter using OTAs and DO-CCII. Int. J. Electron. **94**(2), 99–105 (2007)

9. Horng, J.W., Lee, M.H., Hou, C.L.: Universal active filter using four OTAs and one CCII. Int. J. Electron. **78**(5), 903–906 (1995)

10. Arora, T.S., Sharma, R.K.: Adjoint-KHN equivalent realization of current mode universal biquad employing third generation current conveyor. Indian J. Sci. Technol. **9**(13) (2016)

11. Senani, R., Gupta, M., Bhaskar, D.R., Singh, A.K.: Generation of equivalent forms of operational trans-conductance amplifier-RC sinusoidal oscillators: The nullor approach. J. Eng. **1**(1) (2014)

12. Arora, T.S., Sharma, R.K.: Realization of current-mode KHN-equivalent biquad employing third generation current conveyor. In: IEEE ICCTICT, pp. 20–24 (2016)

# On the Design of Hexagon-Slotted Circular Fractal Antenna for C Band Applications

Harpreet Kaur and Jagtar Singh

**Abstract** A Hexagon-Slotted Circular Fractal Antenna (HSCFA) is demonstrated in this paper. The design of HSCFA is done on FR4 glass epoxy substrate which has thickness of 1.59 mm and relative permittivity of 4.4. Different iterations of HSCFA are performed to improve performance parameters such as radiation pattern, Return Losses (RL), VSWR, Gain (G), etc. Size of the proposed antenna is reduced with the increase in number of iterations. Proposed antenna has return loss of −25.20 dB along with gain of 7.89 dB at the frequency of 6.89 GHz. Coaxial feed technique has been used in feeding the antenna. Antenna size has been reduced than the reference circular antenna. The antenna covers C band for its applications by applying fractal shape in the antenna geometry. The simulation results of antenna are calculated using Ansoft HFSS (high frequency structure simulator) software developed by Ansys Inc. Proposed antenna is fabricated on FR4 substrate material and measured results are calculated experimentally [1]. Simulated and experimental results are good agreement with each other.

**Keywords** Fractal · Hexagonal · FR4 · HFSS · Substrate · VSWR · RL

## 1 Introduction

The word "FRACTAL" derived from the Latin word "fractus" means fragmented. Various natural objects like trees, snowflakes, clouds, etc., are well-known examples of fractal geometries due to their irregular geometries and shapes [2]. Fractal is used to define a family of composite shapes that have self-similarity in their geometrical organization. Fractal antenna becomes a very interesting topic in the area of research

H. Kaur (✉) · J. Singh
Yadavindra College of Engineering, Punjabi University, GKC,
Talwandi Sabo, Bathinda 151302, Punjab, India
e-mail: harpreetghatora6@gmail.com

J. Singh
e-mail: jagtarsivian@gmail.com

© Springer Nature Singapore Pte Ltd. 2018
R.K. Choudhary et al. (eds.), *Advanced Computing
and Communication Technologies*, Advances in Intelligent Systems
and Computing 562, https://doi.org/10.1007/978-981-10-4603-2_29

due to the advantages like light weight, small size, low profile, low cost, and ease in fabrication [3]. It is being used in large range of applications such as satellite communications, radar, mobile communication base stations, missiles, aircraft, and handsets as well as in biomedical telemetry services [4]. A fractal microstrip antenna consist of a dielectric substrate which is sandwiched between ground and microstrip patch planes. However, conventional microstrip patch antennas have some limitations such as poor efficiency, narrow bandwidth, and low gain [5]. To overcome these limitations of antenna, several techniques have been developed. These types of techniques include fractal antenna, slotted patch, and thick substrates with low dielectric constant [6]. Hexagon-Slotted Circular Fractal Antenna is obtained by subtracting hexagonal slots from the circular patch. In this paper, 21 hexagonal slots are subtracted from the patch to study effects on various parameters of the antenna. It enhances the performance of proposed antenna. HSCFA can be used for C band applications. It is excited using coaxial probe feed technique and this technique has numerous advantages over other feeding techniques.

## 2   Antenna Geometry

The proposed antenna is based on fractal geometry shown in Fig. 1. For designing of the suggested antenna, steps are given below:

Step 1: The design of HSCFA is done on FR4 glass epoxy substrate which has thickness of 1.59 mm and relative permittivity of 4.4. Ground plane is a square with side length of 44 mm.

Step 2: A circular patch of radius 16.2 mm placed at the center of the substrate used as a base to construct fractal antenna as shown in Fig. 1a.

Step 3: In first iteration, a hexagonal slot of side 6 mm is subtracted from the circular patch as shown in Fig. 1b.

Step 4: In second iteration four more hexagonal slots having side length one-third of side of hexagon that is 2 mm are subtracted as shown in Fig. 1c.

Step 5: In third iteration, length of the hexagon slots is equal to one-ninth of side of hexagon in first iteration that is 0.67 mm subtracted along the previous slot from the patch as shown in Fig. 1d. Due to fabrication limitations, the countless iterative structure is not possible. Therefore, third iteration circular fractal antenna with twenty one hexagon slots has been finalized. The antenna is fed by coaxial feed technique used for providing the excitation. Outer cylinder of radius 1 mm of material vacuum and inner cylinder of radius 0.5 mm of material pec is used in coaxial feed. This feed technique provides reduced dispersion and less radiation leakage than microstrip transmission lines [7]. This feed technique has another advantage that it can be positioned at any desired location at patch to attain better impedance matching and to evade undesirable radiations from the feed. The dimensions of suggested antenna are given in Table 1.

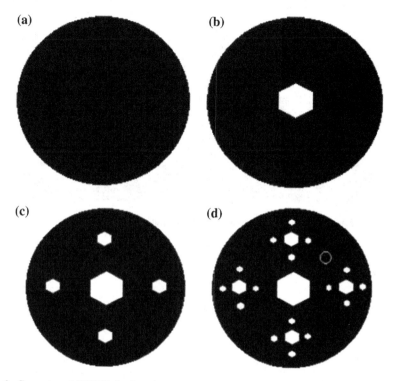

**Fig. 1** Geometry of HSCFA for iterations **a** zero **b** first **c** second and **d** third

**Table 1** Various parameters of the HSCFA

| S.No. | Parameters | Values |
|---|---|---|
| 1 | Substrate material | FR4_Epoxy |
| 2 | Relative permittivity of the substrate | 4.4 |
| 3 | Length of the substrate ($L$) | 44 mm |
| 4 | Width of the substrate ($W$) | 44 mm |
| 5 | Height of the substrate ($H$) | 1.59 mm |
| 6 | Feed type | Coaxial feed |
| 7 | Radius of circular patch | 16.2 mm |
| 8 | Side of largest hexagon slot | 6 mm |
| 9 | Side of larger hexagon slot | 2 mm |
| 10 | Side of smallest hexagon slot | 0.67 mm |

The proposed antenna is also fabricated with FR4 substrate. The different iterations of HSCFA are tested using Vector Network Analyzer (VNA) kit. Simulated and measured results of the fabricated antenna are also compared. Figure 2 reveals with the fabricated top view of HSCFA for second and third iterations. SMA female connector with Teflon inside is used for coaxial feed.

**(a)**                                          **(b)**

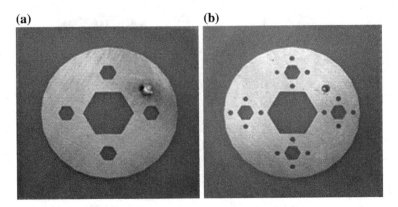

**Fig. 2** Fabricated antennas for **a** second and **b** third iterations

## 3   Results and Discussions

Simulation and experimentally calculated results of HSCFA are described in this section.

### 3.1   Return Loss

Return loss or scattering parameters or S-Parameters graph is used to measure the reflection and transmission losses between the incident and reflected waves [8]. RL comparison of HSCFA for the iterations zero, one, two and three are shown in Fig. 3. This figure clears that all the iterations of proposed antenna are capable of

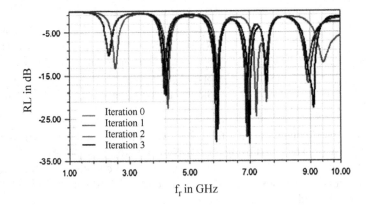

**Fig. 3** Comparison of simulated results of various iterations of proposed antenna (color figure online)

producing multiple resonant frequencies. A comparison of measured and simulated RL versus frequency plot of various iterations of HSCFA is described in Figs. 4 and 5. The VSWR values at given resonant frequencies are between 1 and 2 which are acceptable. Return loss for second iteration at resonant frequencies 4.17, 5.90, and 6.91 GHz are −19.31, −21.01, and −29.31 dB, respectively, whereas experimentally measured return loss for third iteration at resonant frequencies 4.16, 5.88, and 6.89 GHz are −18.50, −31.16, and −25.20 dB, respectively.

## 3.2 Radiation Pattern

This parameter provides the information that how an antenna directs the energy it radiates [9]. Either a rectangular or a polar format is used to present these pattern

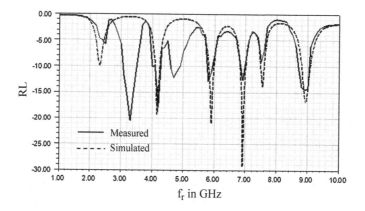

**Fig. 4** Simulated and experimentally measured RL versus $f_r$ of second iteration

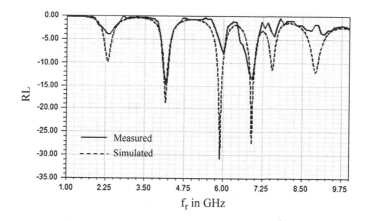

**Fig. 5** Simulated and experimentally measured RL versus $f_r$ of third iteration

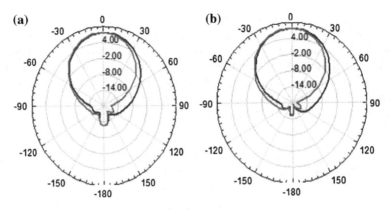

**Fig. 6** Two dimensional radiation pattern of **a** second iteration at 6.91 GHz **b** third iteration at 6.89 GHz

measurements [10]. The two dimensional radiation patterns of second and third iterations of proposed antenna at resonating frequency 6.91 and 6.89 GHz for $\phi = 0°$ and $\phi = 90°$ are displayed in Fig. 6. The gain of proposed antenna is 7.89 dB for third iteration and 7.2 dB for second iteration. Thus with increase in iteration, the number gain of antenna is increased.

### 3.3 VSWR

To get impedance matching, Voltage Standing Wave Ratio (VSWR) of antenna is used [11]. Values of VSWR are 1.16 and 1.53 for resonant frequency of 6.91 and 6.89 GHz. The proposed antenna has acceptable values of VSWR for all the iterations.

The results of proposed antenna in terms of RL and Gain are compared with existing research paper and are shown in Table 2. From this table, it is clear that proposed antenna has more Gain than the existing antenna.

**Table 2** Comparison with other antennas

| Antenna | Operating frequency (GHz) | Return loss (dB) | Gain (dB) |
|---|---|---|---|
| Proposed antenna | 6.89 | −25.20 | 7.89 |
| Nitasha [2] | 6.6 | −18.12 | 6.27 |
| Garima [4] | 7.41 | −34.1 | 5.25 |

# 4 Conclusion

The proposed Hexagon-Slotted Circular Fractal Antenna (HSCFA) has compact size of 44 mm × 44 mm × 1.59 mm having multiband characteristics covering various bands of wireless communications that fulfills the requirements of C band applications. The proposed antenna final iteration has improved gain of 7.89 dB and return loss of −25.20 dB at the resonating frequency of 6.89 GHz. HFSS software is used to analyze the results of proposed antenna. Proposed antenna also has satisfactory values of return loss, VSWR, radiation pattern and other parameters for all the iterations.

# References

1. HFSS (High Frequency Structure Simulator) Software. http://www.ansys.com/Products/Electronics/ANSYS-HFSS
2. Bisht, N., Kumar, P.: A dual band fractal circular microstrip patch antenna for C band applications. In: PIERS Proceedings, Suzhou, China, pp. 852–855. 12 Sept 2011
3. Prasad, K.D.: Antenna and Wave Propagation. Satya Prakashan, New Delhi (2005)
4. Bhatnagar, D., Saini, J.S., Saxena, V.K., Saini, L.M.: Design of broadband circular patch microstrip antenna with diamond shape slot. Int. J. Radio Space Phys. 40, 275–281 (2011)
5. Chitra, R.J., Yoganathan, M., Nagarajan, V.: Coaxial fed double L band microstrip patch antenna array for WiMAX and WLAN application. In: International Conference on Communication and Signal Processing, pp. 1159–1164, 3–5 Apr 2013
6. Singh, A., Mohammad, A., Kamakshi, K., Mishra, A., Ansari, J.A.: Analysis of F-shape microstrip line fed dual band antenna for WLAN applications. In: Springer Conference on Science and Business Media, vol. 20, pp. 133–140, 14 May 2013
7. Balanis, C.A. (ed.): Antenna Theory: Analysis and Design. Wiley, New York (2009)
8. Bhatia, S.S., Sivia, J.S., Kaur, M.: Comparison of feeding techniques for the design of microstrip rectangular patch antenna for X band applications. Int. J. Adv. Technol Eng. Sci. 03, 455–460 (2015)
9. Singh, I., Tripathi, V.S.: Microstrip patch antenna and its applications: a survey. Int. J. Comput. Technol. Appl. 2(5), 1595–1599 (2011)
10. Singh, R., Sappal, A.S., Bhandari, A.S.: Efficiency design of Sierpinski fractal antenna for high frequency applications. Int. J. Eng. Res. Appl. 4, 44–48 (2014)
11. Sivia, J.S., Pharwaha, A.P., Kamal, T.S.: Analysis and design of circular fractal antenna using artificial neural networks. Prog. Electromagn. Res. B 56, 251–267 (2013)

# An Efficient Single-Layer Crossing Based 4-Bit Shift Register Using QCA

Trailokya Nath Sasamal, Ashutosh Kumar Singh
and Umesh Ghanekar

**Abstract** In co-planar QCA fabrication, QCA wire crossing is a challenging task as defects appear to be inherent due to two cell types in single layout structure. In this paper, a compact 2:1 multiplexer is presented which occupies a minimum area in QCA technology. This work also showcases a successful implementation and simulation of 4:1 multiplexer, level trigger D flip-flop and 4-bit shift register using QCADesigner tool. The 4:1 multiplexer and shift register are more robust and enjoy single-layer wire crossing, which requires only one type of cells. The comparison results show superiority of the proposed designs over the previous designs in terms of complexity and delay.

**Keywords** QCA · Shift register · Multiplexer · D flip-flop

## 1 Introduction

Quantum dot cellular automata (QCA) based devices are going to replace CMOS based devices for its potential improvements. Various QCA-based structures are explored in [1–8]. QCA architectures for flip-flops and multiplexers have been presented in [9–15] as these are essential blocks of various digital systems. But most structures are not robust and vulnerable to fabrication defects due to wire crossing between the QCA components. This work presents, a robust 4-bit shift registers using optimal multiplexer and D flip-flop modules.

T.N. Sasamal (✉) · U. Ghanekar
Department of Electronics & Communication, NIT Kurukshetra,
Kurukshetra, India
e-mail: tnsasamal.ece@nitkkr.ac.in

U. Ghanekar
e-mail: ugnitk@nitkkr.ac.in

A.K. Singh
Department of Computer Applications, NIT Kurukshetra, Kurukshetra, India
e-mail: ashutosh@nitkkr.ac.in

© Springer Nature Singapore Pte Ltd. 2018
R.K. Choudhary et al. (eds.), *Advanced Computing
and Communication Technologies*, Advances in Intelligent Systems
and Computing 562, https://doi.org/10.1007/978-981-10-4603-2_30

315

The paper is organized as follows. Section 2 presents a review on QCA. In Sect. 3, optimal design and detailed analysis of basic structures, such as multiplexer and D flip-flop are given, followed by implementation of an efficient 4-bit shift register in Sect. 4. Simulated results and comparative analysis are addressed in Sect. 5. Finally, conclusions are given in Sect. 6.

## 2 Preliminaries

### 2.1 QCA Cell and Gates

A QCA cell is a quantum well, which has four quantum dots located at four corners of a square. Two injected electrons are free to occupy any of the four dots based on Coulomb repulsion among them. These two electrons position at the two corners due to repulsion and yield two possible polarizations as shown in Fig. 1. By applying proper clocks, electrons are able to tunnel through the inter dot barrier by electrons interaction. Any digital circuits can be made of a combination single QCA cell. In QCA circuits, majority gates and inverters are the basic building blocks. Two possible implementations for an inverter are shown in Fig. 1c, d. QCA layout of a 3-input majority gate is depicted in Fig. 1e. A 3-input majority gate can be represented as $MV(a, b, c) = ab + bc + ca$. By fixing one of the inputs to '0' or '1', an AND gate or OR gate can be realized. Input signals can be made available at the output end, by placing QCA cells in serial manner, as shown in Fig. 1b.

### 2.2 QCA Clocking

For proper functioning of QCA circuits, clocks are necessary. This clocking scheme allows the electrons in a cell to arrange themselves by breaking the inter dot barrier.

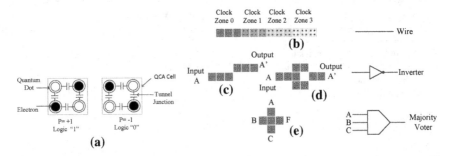

**Fig. 1** **a** QCA. **b** Two polarizations of a cell. **c** QCA wire. **d** Inverters. **e** Majority gate

Generally, it requires four phase clocking scheme. There are four clock zones and each zone is distinct with 90° phase shifted to one another [16]. Each clock zone defines four phases: Switch, Hold, Release, and Relax [17].

## 2.3  QCA Crossing

Till now, there are two different types of crossover are available. These are co-planar and multilayer. In multilayer crossover, multiple layers are used as in CMOS circuit design for interconnection between components as depicted in Fig. 2a. In co-planar crossover scheme, crossing is executed using two different types of cells as shown in Fig. 2b. Another type of co-planar wire crossing is addressed in Shin [18]. In this work, wire crossing based on interference of clocking phases is used, as depicted in Fig. 3.

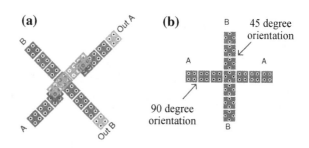

**Fig. 2** Wire crossing. a Multilayer. b Co-planar

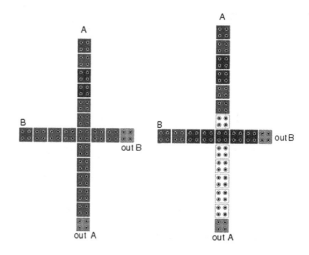

**Fig. 3** Wire crossing using single cell

# 3 Proposed QCA Structures

In this section, we present the basic blocks of a shift register, i.e., multiplexer and D flip-flop using QCA.

## 3.1 2:1 Multiplexer

Figure 4a depicts our multiplexer, which employs 3 MV, and 1 inverter in QCA implementation. This structure comprises only 18 quantum cells. In this structure, MVs in first level implement two AND gates that are driven by first clocking zone. The outputs of these MVs are fed to second level MV which is positioned in the second clocking zone. Select line 'select' is used to select one of the inputs, which also driven by first clock zone. QCA layout for 2:1 multiplexer shown in Fig. 4b provides a valid output after two clock phases. It comprises only 18 cells that spread over an area of 0.016 $\mu m^2$, which has outperformed previous works.

## 3.2 4:1 Multiplexer

Design of 4:1 multiplexer, even larger multiplexer and incorporating with signal distribution network (SDN) is a challenging task. For such systems, the complexity increases in terms of number of wires crossing, so more prone to defects that occurs due to QCA fabrication of single-layer crossing using two different cells (90° and 45°) in a single layout. In this work, we have considered a robust single-layer crossing method using single cell (90°). Figure 5 illustrates the QCA layout for 4:1 multiplexer. It includes three 2:1 multiplexer modules, which has a simple structure that facilitates modularity. Wire crossing is shown by solid squares. Select line $S_0$ is used to select one of the inputs to 2:1 multiplexers at the first level, where $S_1$ is used to select one of the outputs from first level. The proposed circuit results a correct output after eight clock phases and the output cell is driven by forth clock zone.

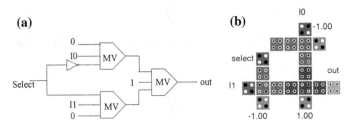

**Fig. 4** 2:1 multiplexer. **a** Schematic. **b** QCA layout

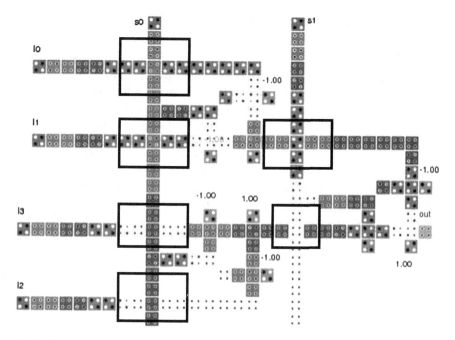

**Fig. 5** QCA layout of 4:1 multiplexer

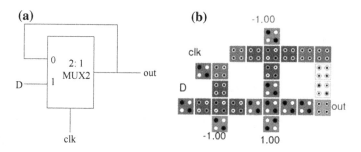

**Fig. 6** 2:1 multiplexer based level trigger D flip-flop. **a** Schematic. **b** QCA layout

## 3.3 D Flip-Flop

Schematic and layout for level sensitive D flip-flop proposed in this paper are demonstrated in Fig. 6a, b, respectively. Majority gate at the second level is driven by third clocking zone and receives inputs from the MVs at the first level. Storing the input data bit in a loop is implemented using four clocking zones, which allows data to be holds in the loop until *clk* = '0'. QCA implementation of level trigger D flip-flop requires 26 cells and area usage is 0.025 μm².

## 4 Proposed 4-Bit Shift Register

Shift registers are the essential circuits in digital system particularly useful in binary division and multiplication with serial and parallel inputs. Figure 7 shows the block diagram of 4-bit shift register, which comprises four D flip-flops and four 2:1 multiplexers. QCA layout of these four edge trigger D flip-flops is represented by solid boxes in Fig. 8 [19]. Serial Input Serial Output (SIPO) and Parallel Input Parallel Output (PIPO) operations are selected by controlling select line of multiplexers, i.e., *shift* = '1' or '0' respectively. Serial data are shifted right and available at output lines after each rising edge of the clock signal. QCA layout of proposed shift register is depicted in Fig. 8. This co-planar implementation involves single-cell wire crossing, which incurs extra delays, but mitigates defects due to manufacturing of two different cells in single QCA layout. It includes 889 cells, spread over an area of 1.4 $\mu m^2$ and the maximum delay is 9 clock cycles.

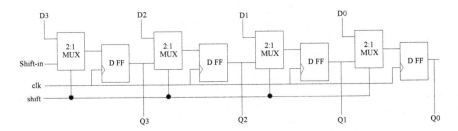

**Fig. 7** Block diagram for 4-bit shift register

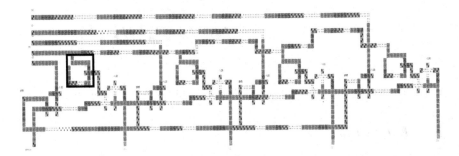

**Fig. 8** QCA implementation of 4-bit shift register

# 5 Simulation Results

Simulator QCADesigner-2.0.3 has been used to verify the designs functionality [20]. The simulation engine used bistable approximation with default parameters values. The simulation results of the 2:1 multiplexer are presented in Fig. 9a. For better readability, we have included only the clock zone 1 at which a valid output is expected. Table 1 summarizes a comparison between considered 2:1 multiplexers. It indicates the superiority of the presented 2:1 multiplexer by achieving a minimal area–delay product (ADP) with lesser complexity.

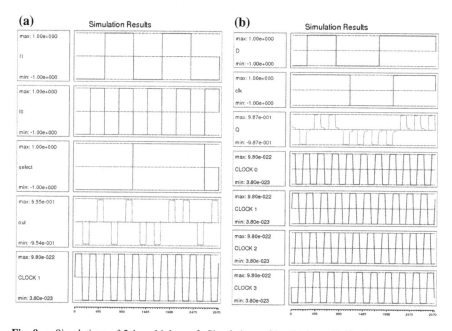

**Fig. 9** **a** Simulations of 2:1 multiplexer. **b** Simulations of level trigger D flip-flop

**Table 1** Comparison of QCA 2:1 multiplexers

| 2:1 multiplexer | Area $(\mu m)^2$ | Cell count | Delay | Clock phases | ADP | Crossover type |
|---|---|---|---|---|---|---|
| [9] | 0.14 | 88 | 1 | 4 | 0.14 | Co-planar |
| [10] | 0.14 | 66 | 1 | 4 | 0.14 | Co-planar |
| [11] | 0.08 | 46 | 1 | 4 | 0.08 | Multilayer |
| [12] | 0.06 | 36 | 1 | 4 | 0.06 | Multilayer |
| [13] | 0.07 | 56 | 1 | 4 | 0.07 | Co-planar |
| [14] | 0.03 | 27 | 0.75 | 3 | 0.022 | Co-planar |
| [15] | 0.02 | 26 | 0.5 | 2 | 0.01 | Co-planar |
| Proposed | 0.016 | 18 | 0.5 | 2 | 0.008 | Co-planar |

**Table 2** Comparison of QCA D flip-flops

| QCA D flip-flop (level trigger) | Area (μm)² | Cell count | Delay | Clock phases | ADP | Crossover type |
|---|---|---|---|---|---|---|
| [5] | 0.08 | 66 | 1.5 | 6 | 0.12 | Co-planar |
| [6] | 0.05 | 48 | 1 | 4 | 0.05 | Co-planar |
| Proposed | 0.025 | 26 | 0.75 | 3 | 0.01 | Co-planar |

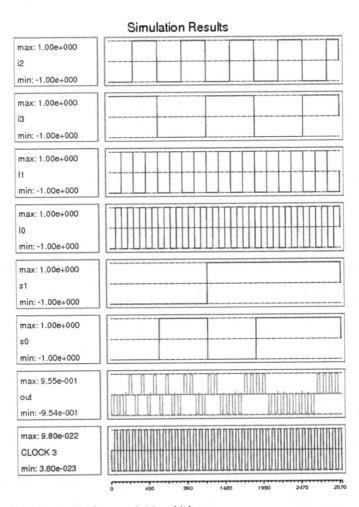

**Fig. 10** Simulation result of proposed 4:1 multiplexer

**Table 3** Comparison of proposed 4:1 multiplexer with existing designs

| 4:1 multiplexer | Area $(\mu m)^2$ | Cell count | Delay | Clock phases | ADP | Crossover type | Maximum QCA line length | Minimum QCA line length |
|---|---|---|---|---|---|---|---|---|
| [13] | 0.24 | 215 | 1.25 | 5 | 0.3 | Co-planar (45°, 90°) | 10 | 2 |
| [14] | – | 94 | 1.75 | 7 | – | Co-planar (45°, 90°) | 15 | 1 |
| [15] | 0.24 | 161 | 3.5 | 14 | 0.84 | Co-planar (45°, 90°) | >30 | 1 |
| Proposed | 0.24 | 169 | 2 | 8 | 0.48 | Co-planar (90°) | 12 | 2 |

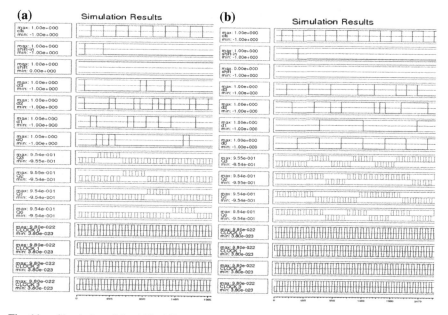

**Fig. 11** **a** Simulation of the 4-bit shift register, when *shift* = '1' (SIPO), **b** Simulation of the 4-bit shift register, when *shift* = '0' (PIPO)

Simulation results of level trigger D flip-flop are illustrated in Fig. 9b. The input and output waveforms confirm the correct operation of the D flip-flop during clock zone 3. Table 2 summarizes a comparative analysis of various D flip-flops, and shows our D flip-flop is better in terms of all design parameters when compared with previous designs [5, 6].

Simulation results of proposed 4:1 multiplexer are depicted in Fig. 10, which validates the correct operation of the design. For instance, when the select lines $S_0$ = '0' and $S_1$ = '0' the multiplexer allows only $I_0$ signal to the output. The output gets the correct input after 2 clock cycles which is more than the previous designs [13, 14], due to the inclusion of single-layer crossing using two cells.

However, the proposed design is less sensitive to defects because of single-cell crossing in single QCA layout. In addition to this, our design is more robust as it maintains minimum and maximum QCA line length of 2 and 12, respectively, in a single clock zone. Table 3 shows a comparison of presented design with prior deigns [13–15] taking all the design parameters and it is worth noticing that the ADP of proposed 4:1 multiplexer is almost close to the existing ones. Serial Input Serial Output (SIPO) operation of 4-bit shift register is verified by making enable signal (*shift*) = '1' as depicted in Fig. 11a. The input signal on *shift-in* shifted from Q3 to Q0 (latched output of D flip-flop) after a delay of 9 clock cycles. When *shift* = '0', the shift register operates in Parallel Input Parallel Output (PIPO) mode and the binary parallel input data are registered as shift register outputs after a delay of 2.25 clock delays shown in Fig. 11b. The proposed 4-bit shift register requires area of 1.4 μm², cell count of 889, and 36 clock phases delay.

# 6 Conclusion

In this work, a robust design for QCA-based multiplexer and shift register is presented using majority gates. Proposed 2:1 multiplexer achieved lesser QCA area and faster than the existing designs. Therefore, an optimal 4:1 multiplexer and a 4-bit shift register have been designed using the proposed 2:1 multiplexer. All the proposed designs are based on co-planar non-crossover wires. Simulations using QCADesigner tool indicate that the presented designs achieve considerable improvements in terms of complexity, area usage, and area-to-delay product.

# References

1. Navi, K., Sayedsalehi, S., Farazkish, R., Azghadi, M.R.: Five-input majority gate, a new device for quantum-dot cellular automata. J. Comput. Theory Nanosci. 7, 1546–1553 (2010)
2. Navi, K., Farazkish, R., Sayedsalehi, S., Azghadi, M.R.: A new quantum-dot cellular automata full-adder. Microelectron. J. 41, 820–826 (2010)
3. Hashemi, S., Tehrani, M., Navi, K.: An efficient quantum-dot cellular automata full-adder. Sci. Res. Essays 7, 177–189 (2012)
4. Sasamal, T.N., Singh, A.K., Ghanekar, U.: Design of non-restoring binary array divider in majority logic-based QCA. Electron. Lett. (2016). doi:10.1049/el.2016.3188
5. Vetteth, A., Walus, K., Dimitrov, V.S., Jullien, G.A.: Quantum-dot cellular automata of flip-flops. ATIPS Laboratory 2500 University Drive, N.W., Calgary, Alberta, Canada T2 N 1N4 (2003)
6. Hashemi, S., Navi, K.: New robust QCA D flip flop and memory structures. Microelectron. J. 43, 929–940 (2012)
7. Sasamal, T.N., Singh, A.K., Mohan, A.: An optimal design of full adder based on 5-input majority gate in coplanar quantum-dot cellular automata. Optik 127(20), 8576–8591 (2016)
8. Sasamal, T.N., Singh, A.K., Mohan, A.: Efficient design of reversible alu in quantum-dot cellular automata. Optik 127(15), 6172–6182 (2016)

9. Kim, K., Wu, K., Karri, R.: The robust QCA adder designs using composable QCA building blocks. IEEE Trans. Comput. Aid. Des. Integr. Circuits Syst. **26**, 76–183 (2007)
10. Mardiri, V., Mizas, C., Fragidis, L., Chatzis, V.: Design and simulation of a QCA 2 to 1 multiplexer. In: The 12th WSEAS International Conference on Computers, Heraklion, Greece, pp. 511–516 (2008)
11. Teodsio, T., Sousa, L.: QCA-LG: A tool for the automatic layout generation of QCA combinational circuits. In: The 25th IEEE Norchip Conference, Aalborg, pp. 1–5 (2007)
12. Hashem, S., Azghadi, M., Zakerol hosseini, A.: A novel QCA multiplexer design. In: The International Symposium on Telecommunications, pp. 692–696 (2008)
13. Mardiris, V.A., Karafyllidis, I.G.: Design and simulation of modular $2^n$ to 1 quantum-dot cellular automata (QCA) multiplexers. Int. J. Circuit Theory Appl. **38**, 771–785 (2010)
14. Roohi, A., Khademolhosseini, H., Sayedsalehi, S., Navi, K.: A novel architecture for quantum-dot cellular automata multiplexer. Int. J. Comput. Sci. **8**, 55–60 (2011)
15. Nadooshan, R.S., Kianpour, M.: A novel QCA implementation of MUX-based universal shift register. J. Comput. Electron. **13**, 198–210 (2014)
16. Wang, Y., Lieberman, M.: Thermodynamic behavior of molecular-scale quantum-dot cellular automata (QCA) wires and logic devices. IEEE Trans. Nanotechnol. **3**, 368–376 (2004)
17. Lent, C.S., Liu, M., Lu, Y.: Bennett clocking of quantum-dot cellular automata and the limits to binary logic scaling. Nanotechnology **17**, 4240–4251 (2006)
18. Shin, S.H., Jeon, J.C., Yoo, K.Y.: Design of wire-crossing technique based on difference of cell state in quantum-dot cellular automata. Int. J. Control Autom. **7**(4), 153–164 (2014)
19. Yang, X., Cai, L., Zhao, X.: Low power dual-edge triggered flip-flop structure in quantum dot cellular automata. Electron. Lett. **46**, 825–826 (2010)
20. Walus, K., Dysart, T.J., Jullien, G.A., Budiman, R.A.: QCA designer: a rapid design and simulation tool for quantum-dot cellular automata. IEEE Trans. Nanotechnol. **3**, 26–31 (2004)

# Wave Digital Realization of Current Conveyor Family

Richa Barsainya, Tarun Kumar Rawat and Rachit Mahendra

**Abstract** The wave digital equivalents of various passive and active elements using wave theory were formulated in yesteryears. Current conveyors have received considerable attention in present scenario. A generalized current conveyor (GCC) represents 12 concrete current conveyors. In this paper, 12 wave digital equivalent circuits of a generalized current conveyor are presented which can further be utilized for the development of a new class of wave digital filters.

**Keywords** Wave digital theory · Digital realization · Current conveyor · Wave equations · GCC

## 1 Introduction

The wave theory was first introduced by Fettweis, by incorporating the concepts of incident and reflected wave quantities [1]. The digital realizations for many passive and active elements like conductances, voltage sources, and wire interconnections have been developed by using wave theory [2, 3]. These developed digital realizations are called wave digital elements, which also preserve the kind of functionality associated with it. Wave digital filters (WDFs) were originated as a way of discretizing an analog reference network by incorporating the concepts of incident and reflected wave quantities of wave theory along with the use of certain discretization procedure. Every WDF is derived from its reference analog filter. The WDFs are obtained by replacing the analog elements with their wave digital

R. Barsainya (✉) · T.K. Rawat · R. Mahendra
Electronics and Communication Division, Netaji Subhas Institute of Technology,
Sector-3, Dwarka, New Delhi 110078, India
e-mail: richa.barsainya@gmail.com

T.K. Rawat
e-mail: tarundsp@gmail.com

R. Mahendra
e-mail: rachitmahendra@yahoo.co.in

© Springer Nature Singapore Pte Ltd. 2018
R.K. Choudhary et al. (eds.), *Advanced Computing
and Communication Technologies*, Advances in Intelligent Systems
and Computing 562, https://doi.org/10.1007/978-981-10-4603-2_31

equivalents [1, 2, 4]. These WDFs not only preserve the structure of the original system, but also its properties like stability or passivity, minimal parameter sensitivity, etc., as it emulates the analog filter. Further wave theory is utilized to derive wave digital equivalent of generalized immittance converter (GIC) leading to development of wave digital filters from an analog reference configuration comprising GICs [5, 6]. A variety of wave digital filters are available and many have found applications in systems requiring digital filters [7, 8]. To design a wave digital filter, it is necessary to acquire the wave digital equivalent of the analog elements. In this paper, wave digital structure of an analog building block, i.e., current conveyor is developed and its wave digital equivalent is presented. The concept of incident and reflected wave of wave theory is utilized to derive the wave digital equivalent of current conveyor family. These digital realizations of current conveyors can be further utilized for development of a new class of wave digital filter based on analog current conveyors.

Current conveyors are used in analog signal processing and were introduced in 1970s and its successive generations were developed thereafter [9]. Presently, there are three generations of current conveyors, namely first, second and third generation (CCI, CCII and CCIII) [10, 11]. Due to their interesting properties like simple circuit, wide bandwidth, low power consumption, and dynamic range, current conveyors are widely used to implement filters as compared to other active elements [12, 13].

In this paper, we have developed the terminal relations for the generalized current conveyor (GCC) [14], which is capable of implementing 12 different current conveyors by appropriate tuning of the coefficients $\alpha$, $\beta$, and $\gamma$, so that their interesting properties can be explored in the digital domain. Our intention in this paper is to present the wave digital realization of the current conveyor using the wave theory. These digital realizations of current conveyors hold all the properties of current conveyor owing to features or realization criteria of wave digital filters.

## 2   Wave Digital Representation of Current Conveyors

A generalized current conveyor (GCC) is a three-port device, whose terminal relations are given below and block representation is shown in Fig. 1.

**Fig. 1** General current conveyor

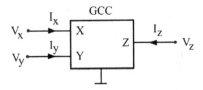

$$\begin{bmatrix} V_x \\ I_y \\ I_z \end{bmatrix} = \begin{bmatrix} 0 & \alpha & 0 \\ \beta & 0 & 0 \\ \gamma & 0 & 0 \end{bmatrix} \begin{bmatrix} I_x \\ V_y \\ V_z \end{bmatrix}$$

This general current conveyor is capable of implementing 12 different current conveyors of first, second, and third generations by proper selection of coefficients $\alpha$, $\beta$, and $\gamma$ for each of the current conveyors. Here, $\alpha$ gives the information about inverting and non-inverting current conveyors. $\alpha = 1/-1$ characterizes an inverting/non-inverting current conveyors. $\beta$ gives the information about the generation of current conveyors. $\beta = 1$ stands for first generation, $\beta = 0$ for the second generation and $\beta = -1$ stands for third generation current conveyors. $\gamma = 1$ is for positive current conveyors whereas $\gamma = -1$ stands for negative current conveyors.

Any analog n-port network can be characterized by applying the concepts of incident and reflected wave quantities given as

$$\left. \begin{array}{l} A_k = V_k + I_k R_k \\ B_k = V_k - I_k R_k \end{array} \right\},$$

where $A_k$ and $B_k$ are referred to as the incident and reflected wave quantities, respectively, and for three-port network, $k = x, y, z$. $R_k$ is the port resistance, $V_k$ and $I_k$ are the voltage and current at the $k$th port, respectively. Wave characterization of GCC is shown in Fig. 2. For a general three-port current conveyor, the following equation holds good

$$A_x = V_x + I_x R_x, \quad B_x = V_x - I_x R_x \tag{2}$$

$$A_y = V_y + I_y R_y, \quad B_y = V_y - I_y R_y \tag{3}$$

$$A_z = V_z + I_z R_z, \quad B_z = V_z - I_z R_z, \tag{4}$$

where $V_x, V_y, V_z$ are the terminal voltages, $I_x, I_y, I_z$ are entering currents, and $R_x, R_y, R_z$ are the resistances offered by the ports $X$, $Y$, and $Z$, respectively. Using above wave equations and terminal relations of GCC, it can be shown that

**Fig. 2** Wave characterization of general current conveyor

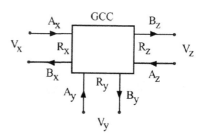

**Table 1** Wave digital equations for realization of first generation current conveyors

| Type | Condition | Final equations |
|---|---|---|
| CCI+ | $\alpha\beta = 1$ or $R_x \neq R_y$ | if $R_y = 2R_x = 2R_z$ |
| | $B_x = \frac{R_y + R_x}{R_y - R_x}A_x - \frac{2R_x}{R_y - R_x}A_y$ | $B_x = 3A_x - 2A_y$ |
| | $B_y = \frac{2R_y}{R_y - R_x}A_x - \frac{R_x + R_y}{R_y - R_x}A_y$ | $B_y = 4A_x - 3A_y$ |
| | $B_z = \frac{-2R_y}{R_y - R_x}A_x + \frac{2R_z}{R_y - R_x}A_y - A_z$ | $B_z = -2A_x + 2A_y - A_z$ |
| CCI− | $\alpha\beta = 1$ or $R_x \neq R_y$ | if $R_y = 2R_x = 2R_z$ |
| | $B_x = \frac{R_y + R_x}{R_y - R_x}A_x - \frac{2R_x}{R_y - R_x}A_y$ | $B_x = 3A_x - 2A_y$ |
| | $B_y = \frac{2R_y}{R_y - R_x}A_x - \frac{R_x + R_y}{R_y - R_x}A_y$ | $B_y = 4A_x - 3A_y$ |
| | $B_z = \frac{2R_y}{R_y - R_x}A_x + \frac{2R_z}{R_y - R_x}A_y - A_z$ | $B_z = 2A_x + 2A_y - A_z$ |
| ICCI+ | $\alpha\beta = -1$ or $R_x \neq -R_y$ | if $R_x = R_y = R_z$ |
| | $B_x = \frac{R_y - R_x}{R_y + R_x}A_x$ | $B_x = 0$ |
| | $B_y = \frac{-2R_y}{R_y - R_x}A_x + \frac{R_x + R_y}{R_y - R_x}A_y$ | $B_y = -A_x$ |
| | $B_z = \frac{2R_z}{R_y - R_x}A_x - A_z$ | $B_z = A_x - A_z$ |
| ICCI− | $\alpha\beta = -1$ or $R_x \neq R_y$ | if $R_x = R_y = R_z$ |
| | $B_x = \frac{-R_x - R_y}{R_y + R_x}A_x$ | $B_x = 0$ |
| | $B_y = \frac{-2R_y}{R_y - R_x}A_x + \frac{R_x - R_y}{R_y + R_x}A_y$ | $B_y = -A_x$ |
| | $B_z = \frac{2R_z}{R_y + R_x}A_x - A_z$ | $B_z = -A_x - A_z$ |

**Table 2** Wave digital equations for realization of second-generation current conveyors

| Type | Condition | Final equations |
|---|---|---|
| CCII+ | $\alpha\beta = 0$ or $R_x \neq 0$ | if $R_x = 2R_z$ |
| | $B_x = -A_x - 2A_y$ | $B_x = -A_x + 2A_y$ |
| | $B_y = A_y$ | $B_y = A_y$ |
| | $B_z = \frac{2R_z}{R_x}(A_x - A_y) - A_z$ | $B_z = A_x - A_y - A_z$ |
| CCII− | $\alpha\beta = 0$ or $R_x \neq 0$ | if $R_x = 2R_z$ |
| | $B_x = -A_x + 2A_y$ | $B_x = -A_x + 2A_y$ |
| | $B_y = A_y$ | $B_y = A_y$ |
| | $B_z = \frac{2R_z}{R_x}(-A_x + A_y) - A_z$ | $B_z = -A_x + A_y - A_z$ |
| ICCII+ | $\alpha\beta = 0$ or $R_x \neq 0$ | if $R_x = 2R_z$ |
| | $B_x = -A_x$ | $B_x = 3A_x$ |
| | $B_y = A_y$ | $B_y = 2A_x + A_y$ |
| | $B_z = \frac{2R_z}{R_x}A_x - A_z$ | $B_z = A_x - A_z$ |
| ICCII− | $\alpha\beta = 0$ or $R_x \neq 0$ | if $R_x = 2R_z$ |
| | $B_x = -A_x$ | $B_x = -A_x$ |
| | $B_y = A_y$ | $B_y = A_y$ |
| | $B_z = \frac{-2R_z}{R_x}A_x - A_z$ | $B_z = -A_x - A_z$ |

**Table 3** Wave digital equations for realization of third-generation current conveyors

| Type | Condition | Final equations |
|---|---|---|
| CCIII+ | $\alpha\beta = -1$ or $R_x \neq -R_y$ | if $R_x = R_y = R_z$ |
| | $B_x = \frac{-(R_x-R_y)}{R_x+R_y}A_x + \frac{2R_x}{R_x+R_y}A_y$ | $B_x = A_y$ |
| | $B_y = \frac{2R_y}{R_x+R_y}A_x - \frac{R_x-R_y}{R_x+R_y}A_y$ | $B_y = A_x$ |
| | $B_z = \frac{2R_z}{R_x+R_y}\left(A_x - A_y\right) - A_z$ | $B_z = A_x - A_y - A_z$ |
| CCIII− | $\alpha\beta = -1$ or $R_x \neq -R_y$ | if $R_x = R_y = R_z$ |
| | $B_x = \frac{-(R_x-R_y)}{R_x+R_y}A_x + \frac{2R_x}{R_x+R_y}A_y$ | $B_x = A_y$ |
| | $B_y = \frac{2R_y}{R_x+R_y}A_x - \frac{R_x-R_y}{R_x+R_y}A_y$ | $B_y = A_x$ |
| | $B_z = \frac{2R_z}{R_x+R_y}\left(-A_x + A_y\right) - A_z$ | $B_z = -A_x + A_y - A_z$ |
| ICCIII+ | $\alpha\beta = -1$ or $R_x \neq -R_y$ | if $R_x = 2R_y = R_z$ |
| | $B_x = \frac{-(R_x+R_y)}{R_x-R_y}A_x$ | $B_x = -3A_x$ |
| | $B_y = \frac{2R_y}{R_x-R_y}A_x + \frac{R_x+R_y}{R_x-R_y}A_y$ | $B_y = 2A_x + 3A_y$ |
| | $B_z = \frac{2R_z}{R_x-R_y}A_x - A_z$ | $B_z = 4A_x - A_z$ |
| ICCIII− | $\alpha\beta = 1$ or $R_x \neq R_y$ | if $R_x = R_y = R_z$ |
| | $B_x = \frac{-(R_x+R_y)}{R_x-R_y}A_x$ | $B_x = -3A_x$ |
| | $B_y = \frac{2R_y}{R_x-R_y}A_x + \frac{R_x+R_y}{R_x-R_y}A_y$ | $B_y = 2A_x + 3A_y$ |
| | $B_z = \frac{-2R_z}{R_x-R_y}A_x - A_z$ | $B_z = -4A_x - A_z$ |

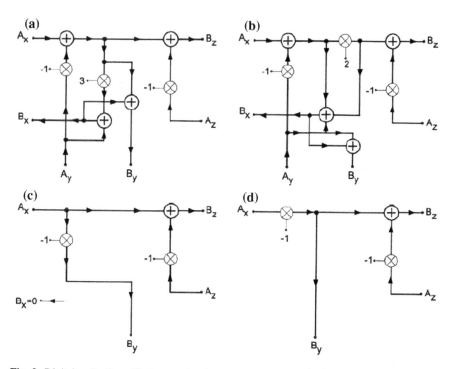

**Fig. 3** Digital realization of first generation current conveyor: **a** CCI+, **b** CCI−, **c** ICCI+, **d** ICCI−

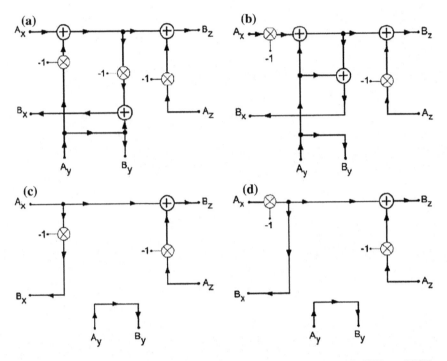

**Fig. 4** Digital realization of second generation current conveyor: **a** CCII+, **b** CCII−, **c** ICCII+, **d** ICCII−

$$(\alpha\beta R_y - R_x)B_x = (\alpha\beta R_y + R_x)A_x - R_x(1+\alpha)A_y \tag{5}$$

$$(\alpha\beta R_y - R_x)B_y = 2\beta R_y A_x - (\alpha\beta R_y + R_x)A_y \tag{6}$$

$$(\alpha\beta R_y - R_x)B_z = -2\gamma R_z A_x + \gamma(1+\alpha)R_z A_y + (R_x - \alpha\beta R_y)A_z \tag{7}$$

By using Eqs. (5–7), relations between reflected wave and incident wave components of the ports X, Y and Z of GCC can be derived. If the condition $(\alpha\beta R_y - R_x) \neq 0$ or $\frac{R_x}{R_y} \neq \alpha\beta$ is satisfied then the wave equations for all the 12 current conveyors can be developed as given in Tables 1, 2 and 3. The digital realizations for the 12 current conveyor configurations are given in Figs. 3, 4 and 5. The wave digital equivalent of current conveyor family or 12 basic current conveyors are efficiently realized.

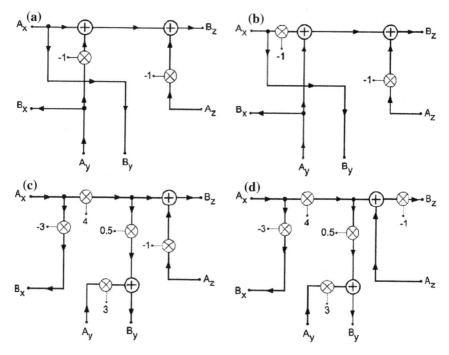

**Fig. 5** Digital realization of third generation current conveyor: **a** CCIII+, **b** CCIII−, **c** ICCIII+, **d** ICCIII−

# 3 Conclusion

In this paper, we have developed wave digital realization for different generations of current conveyors using a general current conveyor which is capable of implementing all members of current conveyor family. These digital realizations can find applications in designing wave digital filters having current conveyor as a building block. The interesting properties of current conveyor in analog domain are also preserved in the digital domain, thereby, giving a large opportunity to designers.

# References

1. Fettweis, A.: Digital filter structures related to classical filter networks. Arch. Elektron. Uebertragungstech. **25**, 19–89 (1971)
2. Fettweis, A.: Wave digital filters: theory and practice. IEEE Proc. **74**, 270–327 (1986)
3. Antoniou, A., Rezk, M.G.: Digital-filter synthesis using concept of generalised-immittance converter. IEEE Proc. Electron. Circuits Syst. **1**(6), 207–216 (1977)
4. Barsainya, R., Rawat, T.K., Kumar, M.: Design of minimum multiplier fractional order differentiator based on lattice wave digital filter. ISA Transactions (2016). In press. doi:10. 1016/j.isatra.2016.09.024, ISSN 00190578

5. Smith, K.C., Sedra, A.: The current conveyor: a new circuit building block. IEEE Proc. **56**, 1368–1369 (1968)
6. Sedra, A., Smith, K.C.: A second-generation current conveyor and its application. IEEE Trans. Circuit Theory **17**, 132–134 (1970)
7. Koton, J., Herencsr, N., Vrba, K.: KHN-equivalent voltage-mode filters using universal voltage conveyors. AEU-Int. J. Electron. Commun. **65**(2), 154–160 (2011)
8. Biolek, D., Vrba, K., Căjka, J., Dostàl, T.: General three port current conveyor: a useful tool for network design. J. Electr. Eng. –Bratislava, **51**, 36–39 (2000)
9. Barsainya, R., Rawat, T.K.: Novel wave digital equivalents of passive elements. In: Annual IEEE India Conference (INDICON-2015), New Delhi, 1–6 (2015). doi:10.1109/indicon.2015.7443461
10. Barsainya, R., Aggarwal, M., Rawat, T.K.: Minimum multiplier implementation of a comb filter using lattice wave digital filter. In: Annual IEEE India Conference (INDICON-2015), New Delhi, 1–6 (2015). doi:10.1109/indicon.2015.7443491
11. Barsainya, R., Rawat, T.K., Mahendra, R.: A new realization of wave digital filters using GIC and fractional bilinear transform. Eng. Sci. Technol. Int. J. **19**(2), 429–437 (2016)
12. Barsainya, R., Aggarwal, M., Rawat, T.K.: Design and implementation of fractional order integrator with reduced hardware. In IEEE International Conference on Signal Processing and Integrated Networks (SPIN-2016), 580–585 (2016)
13. Fabre, A.: Third-generation current conveyor: a new helpful active element. Electron. Lett. **31**, 338–339 (1995)
14. Altuntas, E., Toker, A.: Realization of voltage and current mode KHN biquads using CCCIIs. AEU-Int. J. Electron. Commun. **56**(1), 45–49 (2002)

# Author Index

Printed in the United States
By Bookmasters